The PHYSICS *of* CHRISTIANITY

Frank J. Tipler

DOUBLEDAY
New York London Toronto Sydney Auckland

DD
DOUBLEDAY

Published in the United States by Doubleday, an imprint
of The Doubleday Publishing Group,
a division of Random House, Inc., New York.
www.doubleday.com

A hardcover edition of this book was originally published
in 2007 by Doubleday.

DOUBLEDAY is a registered trademark and the DD colophon is a
trademark of Random House, Inc.

A list of art credits appears on p. 321.

Book design by Michael Collica

Library of Congress Cataloging-in-Publication Data
Tipler, Frank J.
The physics of Christianity / Frank J. Tipler.
p. cm.
Includes bibliographical references.
1. Physics—Religious aspects—Christianity.
2. Religion and science. I. Title.
BL265.P4T57 2007
261.5'5—dc22
2006039028

ISBN 978-0-385-51425-5

First Paperback Edition

146086900

To God's Chosen People, the Jews,
who for the first time in 2,000 years
are advancing Christianity

I will bless those who bless you, and he who curses you, I will curse; and through you will be blessed all the families of the Earth.

—Genesis 12:3

Contents

The PHYSICS *of* CHRISTIANITY

I

Introduction: Christianity as Physics

THE LATEST OBSERVATIONS OF THE COSMIC BACKGROUND radiation show that the universe began 13.7 billion years ago at the Singularity. Stephen Hawking proved mathematically that the Singularity is not in time or in space, but outside both. In other words, the Singularity is transcendent to space and time. According to the theologian Thomas Aquinas, "God created the Universe" means simply that all causal chains begin in God. God is the Uncaused Cause. In physics, all causal chains begin in the Singularity. The Singularity itself has no cause. For a thousand years and more, Christian theologians have asserted that there is one and only one "achieved" (actually existing) infinity, and that infinity is God. The Cosmological Singularity is an achieved infinity.

The Cosmological Singularity is God.

"But," the average person may protest, "the 'Cosmological Singularity' is not *my* idea of God. I picture God as a kindly, white-haired old man, loving but with immense power. The 'Cosmological Singularity' (whatever that is) is too abstract, too intellectual to be *my* God, the God I pray to every night. It sounds like some crazy idea some physicist would dream up. It's definitely not the God of Judaism or Christianity."

Not so. The Cosmological Singularity is the Judeo-Christian God. Think of it this way. Everybody knows that when you flip a light switch, the light goes on because an electrical current flows in the wires in the walls. Everybody also knows that electrons carry the electric charge whose motion makes the electric current. I invite you to imagine an "electron"—you must have *some* image of an electron, since you use the word.

Now let me ask you: when you imagined an "electron," did you imagine an excitation of a quantized, relativistic fermion field, part of an electroweak doublet? Unless you are a professional physicist, I know you didn't. You probably imagined a little ball of some sort. Such an image is good for some purposes, even in physics. One can compute a fairly accurate value for the "drift velocity" of the electrons through the wire using the "little ball" image of the electron. But did you know that the electrons which carry the current in the wire are at a temperature of 80,000 degrees Celsius (140,000 degrees Fahrenheit)?[1] You might wonder, If the conduction electrons are at that high a temperature, why don't they melt the wires? Why don't they start a fire and burn the house down? The reason is that the conduction electrons can't give up their high-temperature energy to the wires. But to understand *why* the electrons can't give up their energy, one has to go beyond the "little ball" image of the electron. (One has to think "quantized fermion.")

Similarly, everyone has an image of "God," but to really understand what God really is and how He could interact with the universe, one must use a theory beyond everyday commonsense physics. Contrary to what many physicists have claimed in the popular press, we have had a Theory of Everything for about thirty years. Most physicists dislike this Theory of Everything because it requires the universe to begin in a singularity. That is, they dislike it because the theory is consistent only if God exists, and most contemporary scientists are atheists. They don't want God to exist, and if keeping God out of science requires rejecting physical laws, well, so be it.

My approach to reality is different. I believe that we have to accept the implications of physical law, whatever these implications are. If they imply the existence of God, well then, God exists.

We can also use the physical laws to tell us what the Cosmological Singularity—God—is like. The laws of physics tell us that our universe began in an initial singularity, and it will end in a final singularity. The laws also tell us that ours is but one of an infinite number of universes, all of which begin and end in a singularity. If we look carefully at the collection of all the universes—this collection is called the *multiverse*—we see that there is a third singularity, at which the multiverse began.

But physics shows us that these three apparently distinct singularities are actually one singularity. The Three are One.

There is one religion which claims that God is a Trinity: Christianity. According to Christianity, God consists of Three Persons: God the Father (the First Person), God the Son (the Second Person), and God the Holy Ghost (the Third Person). But there are not three Gods, only one God. Using physics to study the structure of the Cosmological Singularity, we can see that indeed the three "parts" of the Singularity can be distinguished by employing the idea of personhood. In particular, physics can be used to show how it is possible for a man—Jesus, according to Christianity—to actually *be* the part of the Singularity that connects the Initial and Final Singularities. So the Incarnation makes perfectly good sense from the point of view of physics.

Traditional Christianity has always claimed that "miracles" do not violate ultimate physical law, although a miracle may violate our limited knowledge of physical law. Thus, if we know ultimate physical law—and if our Theory of Everything is correct, we do—we should be able to explain all the miracles of Christianity.

And so we can. The miracle of the Star of Bethlehem was a supernova in the Andromeda Galaxy. The miracle of the Virgin Birth of Jesus, the virgin birth of a male, is plausible if we use modern knowledge of exactly how DNA codes for gender. One expects that, in a virgin birth, all the DNA in the child would come from the mother alone. This is possible if Jesus were an XX male. In the U.S. population, 1 male in 20,000 is an XX male. Using modern DNA technology, it is a simple matter to test whether a male is an XX male. A DNA test was performed on the Shroud of Turin, claimed to be the burial shroud of Jesus, and the Oviedo Cloth, claimed to be the "napkin" that covered Jesus' face in his tomb. The DNA on both relics is just what one would expect if it were the DNA of an XX male.

According to Christians, Jesus rose from the dead in a "resurrection body," a body that we will all have at the Universal Resurrection in the future. This "Glorified Body" was capable of "dematerializing" at one location and "materializing" in another. Modern particle physics provides a mechanism for dematerialization: conversion of the matter of an object into neutrinos, which are elementary particles that interact very weakly with normal matter and thus would be invisible. Reversing the process would result in apparently materializing out of nothing. If this was the mechanism of Jesus' Resurrection, there are several tests that could demonstrate it. In fact, some of these tests are so simple that an ordinary person could carry them out. The image of Jesus on the Turin Shroud has certain features we would expect to arise in the neutrino dematerialization process.

Christians claim that Jesus will come again, at the end of human history. Two developments in physics suggest that human history will end in about fifty years: computer experts predict that computers will exceed human intelligence within fifty years, and the dematerialization mechanism can be used to make weapons that are to atomic bombs as atomic bombs are to spitballs. Such weapons and superhuman computers would make human survival unlikely, and in his discussion of the Second Coming, Jesus said he would return when humans faced a "Great Tribulation" of such magnitude that we would not survive without his direct intervention. We will face such a Great Tribulation within fifty years.

From the perspective of the latest physical theories, Christianity is not a mere religion but an experimentally testable science.

II

A Brief Outline of Modern Physics

The Many-Worlds Interpretation is trivially true.

STEPHEN W. HAWKING[1]

The [Many-Worlds Interpretation] is okay.

MURRAY GELL-MANN, PHYSICS NOBEL LAUREATE[2]

The final approach [to quantum mechanics] is to take the Schrödinger equation seriously, to give up the dualism of the Copenhagen interpretation, and to try to explain its successful rules through a description of measurer and their apparatus in terms of the same deterministic evolution of the wave function that governs everything else. . . . For what it is worth, I prefer this last approach.

STEVEN WEINBERG, PHYSICS NOBEL LAUREATE[3]

I question whether quantum mechanics is the complete and ultimate truth about the physical universe. In particular, I question whether the superposition principle can be extrapolated to the macroscopic level in the way required to generate the quantum measurement paradox. . . . I simply cannot convince myself that any of the solutions proffered to the quantum measurement paradox is *philosophically* [my emphasis] satisfactory.

ANTHONY LEGGETT, PHYSICS NOBEL LAUREATE[4]

I'm afraid I do [believe in the Many-Worlds Interpretation]. I agree with John Wheeler who once said that is too much *philo-*

sophical [my emphasis] baggage to carry around, but I can't see how to avoid carrying that baggage.

PHILIP ANDERSON, PHYSICS NOBEL LAUREATE[5]

I think we are forced to accept the Many-Worlds Interpretation if quantum mechanics is true.

RICHARD P. FEYNMAN, PHYSICS NOBEL LAUREATE[6]

I don't see any way to avoid the Many-Worlds Interpretation, but I wish someone would discover a way out.

LEON LEDERMAN, PHYSICS NOBEL LAUREATE[7]

Jesus answered, "My kingdom is not of this world."

JOHN 18:36

MODERN PHYSICS IS BASED ON THREE FUNDAMENTAL theories: quantum mechanics, general relativity, and the Standard Model of particle physics. In the popular press—and even in many technical physics journals—one will find much discussion of other theories, for example, inflation cosmology, superstring theory, and M-theory. Ignore these other theories. They have no experimental support whatsoever. In contrast, quantum mechanics, general relativity, and the Standard Model have enormous support from experiment. All three theories have made predictions again and again over many decades, predictions that are completely counterintuitive to scientists and the average person, and all of these counter-to-common-sense predictions have been confirmed by experiment. A scientist, if he wishes to remain a scientist, must accept the results of experiment, and nothing but the results of experiment.

Unfortunately, many scientists, even many very good scientists, have a tendency to reject the firmly established physical laws once they realize that these laws have implications which are contrary to the intuitive picture of the world which these scientists formed in childhood. When any scientist rejects the implications of physical law, for any reason other than experiment, then he ceases to be a scientist. He becomes a philosopher, practicing a discipline in which he has no special expertise. When he rejects the implications of physical law without experimental warrant, he is no longer speaking as a scientist; he is speaking as a layman, with no more authority than the average person in the street.

Fortunately, when a scientist leaves the discipline in which his expertise rests for philosophy, he generally retains his scientific habits of honesty. If pressed, he will tell you that he is no longer speaking as a scientist but as a philosopher. Just ask him what the experimental evidence is for his claim, any claim. He will generally tell you that there is none. Any scientist can cite at length the experimental evidence for a true scientific claim.

This will also apply to me. I could talk for hours on the experiments that indicate the truth of quantum mechanics, general relativity, and the Standard Model. Any physicist could, even in those all too common cases when a particular physicist has decided on philosophical grounds that there must be something wrong with one or more of these fundamental theories. Just ask any physicist for the experimental evidence for any of these theories, or consult the physics textbooks. I am therefore not going to waste any space defending the truth of these three fundamental theories of modern physics; I am just going to outline what these theories assert about the nature of physical reality. I'm going to assume that all three theories are actually true. Once again, there is no experiment at all that even suggests otherwise.

Quantum Mechanics

Of the three, quantum mechanics is the most fundamental theory, and also the most counter to everyday intuition about how the physical world operates. Quantum mechanics asserts that every object in the universe—an electron, a chair, you and me, the planet Earth, and the entire universe itself—is simultaneously both a particle and a wave. Unfortunately, our daily experiences cause us to think that the categories of "particle" and "wave" are mutually exclusive, and what makes the theory of quantum mechanics so counterintuitive is its claim that actually everything is both. Even physicists, who know there is overwhelming evidence that everything is simultaneously both a particle and a wave, find it hard to understand. Let me try to explain how this is possible.

A particle is easy to imagine: a ball flying through the air is an excellent model for all particles. A good image for a wave is a wave in the sea, coming in toward the shore. One obvious difference between particles and waves is the fact that the former are localized in space, whereas the latter are spread out in space. But there is a more basic difference: two or more waves can *interfere* with one another, and interfere either constructively or destructively. As we will see, it is this phenomenon of interference that is crucial to understanding quantum mechanics.

Constructive interference between two waves is illustrated in Figure 2.1.

In this figure, two waves—think of two waves moving on the surface of the sea—are moving toward each other. When the waves overlap, the total height of the water is the sum of the height of each wave separately. Imagine that, when the waves overlap, one wave first raises the sea level to its height, and the other wave then raises the raised sea level to *its*

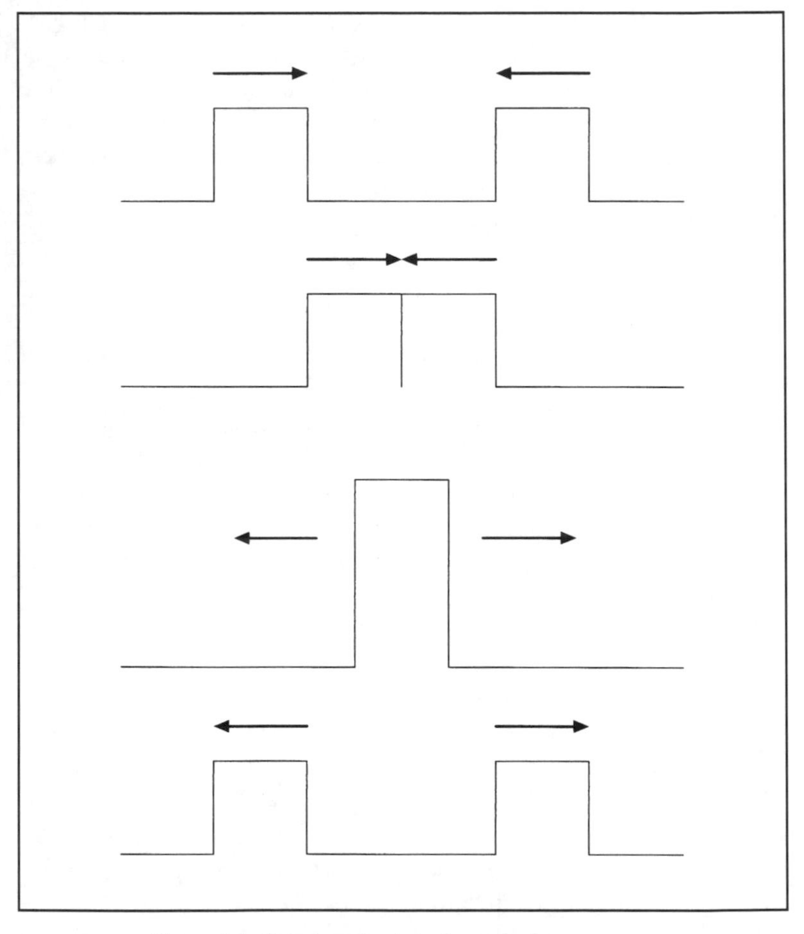

Figure 2.1. Constructive interference of two waves.

height. In the figure, each wave is idealized as a square 2 meters in height and 2 meters in length. So when the waves overlap, the total height of the single wave above the average sea level is 2 + 2, or 4 meters. The adding of the heights of the two waves is called *constructive* because the two heights add. Furthermore, the waves pass through each other, each wave having no residual effect on the other. So *interference* is something of a misnomer, because actually the two waves never permanently add or subtract anything from each other. Since the heights simply add to give the total height of the wave when they overlap (rather than having the total height be the product of the two heights separately, for instance), we say that waves obey the Principle of Linear Superposition (*linear* means "simply add").

Destructive interference between two waves is illustrated in Figure 2.2. As in the previous figure, two waves are moving toward each other, but this time one wave is not a mass of water raised above the average sea level but is instead a depression, a trough below the average sea level. Since the height of the second wave is *below* the average sea level, we say that its height is *negative.* The Principle of Linear Superposition still applies; as before, the heights add, but this time one of the heights is negative. In the figure, one wave is idealized as a square 2 meters in height, while the other is a square trough of height *minus* 2 meters. The total height of the water is thus $2 + (-2) = 0$ meters. In other words, the waves (for an instant) completely cancel—destroy—each other. We have destructive interference. Keep in mind both forms of interference as we now consider how to combine the properties of particles and waves.

Let us first imagine putting a particle on a wave. Imagine, for example, a surfer riding a surfboard on the top of a wave moving in to shore. The top of the wave is actually extended in space, forming a wave *"front."* We can easily picture several surfers riding the same single wave front moving in to shore. An equation for the wave motion would in this case also be an equation for the motion of the surfers. If we know the motion of the wave, an additional equation for the motion of the surfers would be redundant.

An equation for particle motion in terms of a equation for waves that carry the particle was written down in the early part of the nineteenth century: it is called the *Hamilton-Jacobi equation.*[8] By the end of the nineteenth century, the Hamilton-Jacobi (H-J) equation was considered to be the most fundamental and powerful formulation of Newtonian

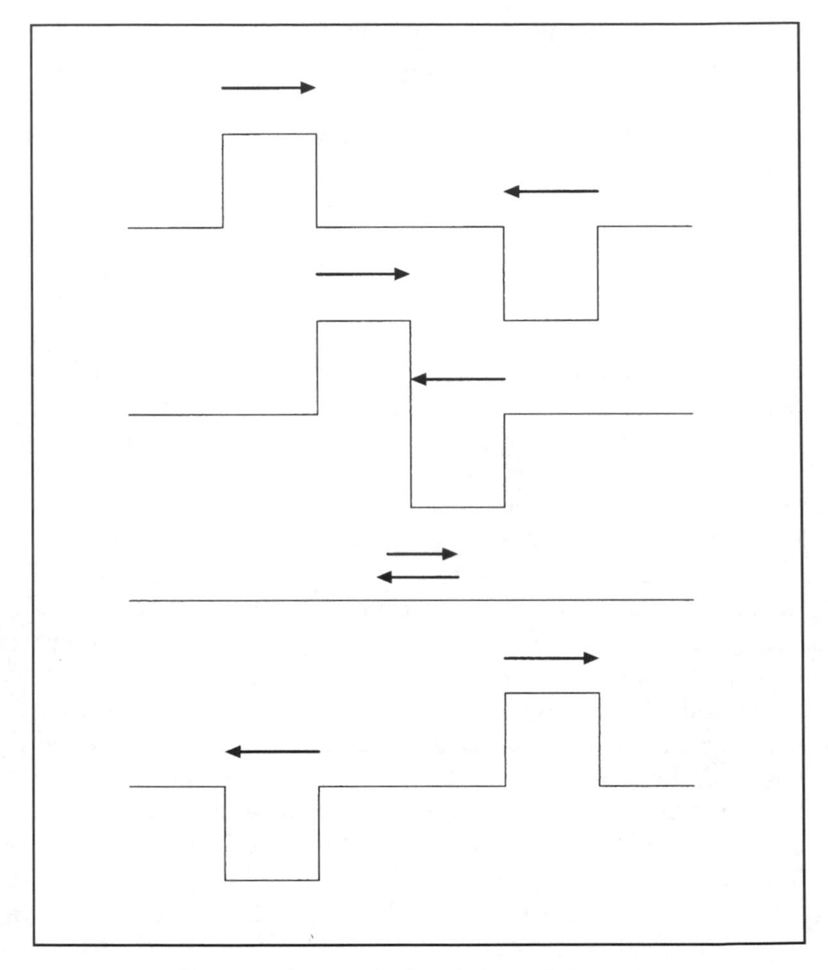

Figure 2.2. Destructive interference of two waves.

mechanics. Unfortunately, the H-J equation had one grave defect: it was nonlinear and claimed the waves developed singularities in a short time.

Imagine a wave on the surface of the sea moving toward a rock in the sea. The wave cannot pass through the rock and so must bend around it. Now imagine two surfers moving on top of the wave, one passing south of the rock and the other north. The part of the wave to the north of the rock would be bent south, carrying the northernmost surfer with it, while the part of the wave to the south of the rock would be bent north, carrying the southernmost surfer with it. The waves—and the two surfers—would collide somewhere beyond the rock.

This example illustrates what would happen to solutions to the H-J equation with an *attractive potential,* such as the gravitational field of the Earth. According to the H-J equation, in the collision the waves would not linearly superpose either constructively or destructively. The H-J is not a normal wave equation with linear superposition. It is nonlinear, which means that the waves cannot pass through each other. Instead, they truly destroy each other: the wave motion at the point of collision is no longer controlled by the H-J equation: the two surfers hit each other with infinite speed. This infinity is the singularity.

In the H-J equation, the predicted singularities would manifest themselves in the laboratory. We shall encounter the term *singularity* many times in this book. A singularity is a place where the equation ceases to apply, usually because some quantity in the equation has become infinite. A singularity occurring in the laboratory would contradict observation: infinite physical quantities have never been observed. If singularities occur, they must occur outside the laboratory, outside of space and time altogether.

The Austrian physicist Erwin Schrödinger solved the Hamilton-Jacobi singularity problem in 1926. In effect, Schrödinger showed that if a *quantum potential* that itself obeyed a certain equation was added to the usual potential of the H-J equation, the two equations were mathematically equivalent to a single equation—now known as Schrödinger's equation—which was linear and which therefore had no singularities. The waves bending around the rock would superpose—and the surfers would pass through each other! As an added benefit, Schrödinger's equation correctly describes the behavior of electrons in atoms. More generally, it has been found to describe correctly the interactions of even large numbers of atoms. It is the fundamental equation of what is now called quantum mechanics. But although the mathematical problem is solved, the problem of interpreting the physical meaning of Schrödinger's wave function remains. What in particular does it mean to say "the surfers pass through each other" when they collide? Why do we not see the wave associated with the particle?

We solve this problem by studying the behavior of the wave function in actual physical situations. Let us take the second question first. Why do we not see the wave but see only the particle? This question was answered by the German physicist Werner Heisenberg in a famous series of lectures he gave at the University of Chicago in the late 1920s.[9] Heisenberg imagined a plane wave moving toward a rectangular array of detectors. He pictured the detectors to be an array of silver halide atomic complexes (such complexes are the active chemical compound in traditional photographic film) or some other sort of detector that would tell us if a moving particle moved through it.

In our surfing model, let us suppose the array consists of a series of concrete columns, each reaching above the average sea level. Suppose these columns form a regular rectangular array: imagine that they are located 10 meters apart in all directions on the surface. That is, if we are on one column, there is another 10 meters to the north, another 10 meters to the south, another 10 meters to the east, and another 10 meters to the west. Let us now imagine that on the top of each column there is a chemical that changes from blue to red if it gets wet.

We now have a detector for wave motion: if a wave of sufficient height passes through the array of columns, the tops of the columns will change from blue to red. Looking down on the array from an airplane, we would see an array of blue dots if no wave has passed. A sufficiently high wave passing through the array of columns would be seen from the airplane as a changing array of colored dots: red on the side where the wave has passed, and blue on the side where it has not yet passed. At any instant, the location of the wave is the location between the blue and the red dots. We will imagine that the array of columns begins somewhere in the water to the east and continues in to the shore, which lies somewhere far to the west. The array we will imagine continues north and south as far as the eye can see.

Heisenberg investigated the effect of a plane wave corresponding to an electron moving through the array, and he showed that if the wave happened to overlap just one of the columns in the first array, say because the wave happened to be slightly higher there, so that that column's top would turn red, then constructive interference would cause the water immediately to the east to be much higher than the part of the wave anywhere else. The result would be that the sea level would rise over the columns in a straight line leading due east from the first column whose top was overlapped. From an airplane looking down, we would see not a wave coming in to shore with a line between the blue and the red but instead a single red line passing through the blue. In other words, we would see a *particle* and not a wave!

This explanation does not completely solve the problem of why we see only a particle, because we have not accounted for only one column being overlapped. It seems possible that a wave whose height was the same along the first line of columns would overlap either all the columns or none of them. And in fact Heisenberg did not answer this objection. He was able to prove only that if just one column in the initial array was overlapped, only the columns immediately to the east would also be overlapped. And Heisenberg assumed in his calculation that the height (amplitude) of the wave was the same for all columns (detectors) in the first line of the array.

The complete solution of why we see only a particle track rather than a wave track was first obtained by a physics graduate student, Hugh Everett, in 1957.[10] Everett pointed out that we are also subject to Schrödinger's equation, which means that we are also both particles and waves. Our wave function is subject to linear superposition, just as the wave functions of electrons and water molecules are. So if we really want to determine what we will actually observe, we have to take into account our quantum mechanical nature also. We can't just suppose the electrons and collections of atoms obey Schrödinger's equation and we don't. After all, we are nothing but large collections of atoms and electrons.

The key idea is to apply linear superposition not only to electrons and atoms but also to us. Suppose that, rather than having an array of columns or detectors, each of us had a single line of columns-detectors from east to west. If a wave were to move from east to west, we would obviously either see nothing (the wave didn't have sufficient height to overlap any column, or trigger any detector) or, if one column was overlapped (initial detector triggered), the entire single row of columns would be overlapped or detectors triggered.

Now Everett noticed the crucial point: we can determine what would happen to the entire array by linear superposition of all the rows of columns. If we superpose, we find necessarily that all are overlapped (or triggered). *But we don't see them all overlapped or triggered because our sensory apparatuses are designed to see only one!* That is, if in fact only one line is overlapped or triggered, our sensory apparatuses—our eyes, our ears, our touch, and so forth—had better perceive only one line. If in fact only one line is triggered, our senses, if our senses and brains are working correctly, had better perceive only one line. But Everett pointed out that linear superposition says that, even if the others are also triggered, we cannot see these other columns being triggered, we can see only one. Nevertheless, quantum mechanics says these other lines of triggered columns are present in reality. And they are seen. They are seen by analogues of ourselves in parallel universes.

This conclusion is termed the *many-worlds interpretation of quantum mechanics.* However, *interpretation* is a misnomer, because it is the *only* interpretation of quantum mechanics. As Everett emphasized, the many worlds, which is to say, the other universes with analogues of ourselves, must necessarily exist if linear superposition applies not only to electrons and atoms and collections of atoms—and innumerable experiments show that it does—but also to those particular collections of atoms called human beings. We are no exception: the physical laws apply to everything.

As the quotations with which I began this chapter show, even Nobel Prize–winning physicists have trouble accepting the many-universes implication of quantum mechanics, or, more precisely, the linear superposition property of quantum mechanics. But make no mistake: if quantum mechanics is true, the many universes necessarily exist. The mathematics of quantum mechanics gives no alternative. The existence of the many universes, which collectively are called the *multiverse,* is really also implied by the Hamilton-Jacobi equation, but because they are nonlinear, one could have supposed that only one particle trajectory was actually followed. The linearity of Schrödinger's equation does not leave us that option. So the multiverse exists even in classical Newtonian mechanics if this theory is expressed in its most powerful mathematical form.

The multiverse is as revolutionary a concept as the idea that the Earth is not the center of the universe but instead merely the third planet from the Sun. In fact, many of the very same objections leveled at the Copernican theory nearly 500 years ago are now being leveled at the multiverse theory. For example, people who do not want to believe in the multiverse argue that the huge increase in the size of reality—a multiverse composed of an infinite number of universes rather than a single universe—violates Occam's (or Ockham's) razor, a principle often invoked in science. A medieval theologian and philosopher, William of Ockham (1285–1349), wrote concerning acceptable theoretical premises: *Pluralitas non est ponenda sine necessitate,* in English, "Plurality must not be postulated without necessity." Indeed, the multiverse is about as great an extension of the plurality of worlds as is possible.

The multiverse does not quite involve all logically possible universes;

it involves only those that are consistent with the laws of physics. There is not, for example, a universe in the multiverse in which magic is allowed. Still, it must be admitted that reality is enormously expanded if in fact the multiverse exists. However, it must be kept firmly in mind that we are not postulating the existence of the multiverse. Instead, we are postulating that quantum mechanics—and classical mechanics in Hamilton-Jacobi form—applies to all systems without exception. Then it follows, of mathematical necessity, that the multiverse exists. Once again, all experiments conducted to date show that quantum mechanics (or classical mechanics) applies to every system we have been able to test over the last century (the last three centuries if we include classical mechanics). The multiverse is forced on us by observation.

Exactly the same Occam's razor argument was used against the Sun-centered solar system when Nicolaus Copernicus first proposed it, in 1543. Before Copernicus (1473–1543), people thought they lived in a rather small, cozy universe, ending at the fixed stars, which themselves were not too far from the Earth. However, it was instantly realized (and even pointed out by Copernicus) that if the Earth were not the center of the universe but instead the third planet from the Sun, which was itself at the center of the solar system, then the stars had to be gigantically much farther away than everyone had previously believed. If the Earth moves around the Sun, then at different times of the year we on the moving Earth will see the stars from different positions, and if the stars are close by, they should appear to shift their positions. The apparent shift is called *parallax,* and no such parallax shift is visible to the unaided eye (stellar parallax was not seen until the early nineteenth century). Therefore, many scholars in the sixteenth century concluded, the Copernican theory could not be true, because it multiplies the amount of space between the stars by an enormous factor. What is the point, they asked, of all that useless space? By Occam's razor, the Copernican theory is multiplying space—the size of reality—without necessity.[11]

There was a necessity, the same necessity that forces the multiverse upon us: to have one set of physical laws for both the small and the large. In the pre-Copernican universe, there was one set of physical laws for the small—the region near the Earth, called the *sublunar region*—and another set for the large—the planets, the Moon, the Sun, and the stars. In fact, scholars before Copernicus believed that things on Earth were composed of fundamentally different substances, namely the four elements—earth, air, fire, and water—than the objects in the heavens, which were composed of the quintessence, which just means "fifth element." The four elements of the Earth obeyed a completely different set of laws than did the elements making up the heavenly bodies. The Copernican Revolution says this is false: all reality obeys one, and only one, set of laws. Similarly, asserting that all reality, not just the small world of atoms and electrons but also the medium-size world of everyday life and the large world of the stars and the universe, obeys quantum mechanics forces us to accept the multiverse. This is a mathematical fact. Denying it is the same as denying that 2 + 2 = 4.

The existence of a multiverse of universes, of which we can see only one universe, means that we can never get sufficient information to determine what will actually happen in the future of the particular universe we see ourselves to be in. We can use quantum mechanics to calculate only the *probability* that a certain event will occur. Probability is *always* an expression of the human limitation of knowledge and is *never* some facet of nature. Thus, probability in quantum mechanics is an expression of human ignorance, but quantum mechanics also says that it is impossible, even in principle, to overcome this ignorance. Furthermore, as we shall see later, quantum mechanics allows us to compute in many cases exactly what these fundamental limitations to our knowledge are.

Let us review an old calculation of a probability, the probability that a die (singular of *dice)* will land with a 5 faceup. As far as we know, the die is an honest die, not weighted to favor any one of its six faces. We also know of no force that will give any preference to any particular face, nor are we aware of anything in the manner in which we plan to toss the die that would give a preference to any face. In fact, there may actually be some reason or several reasons unknown to us why one face of the die is favored. This doesn't matter; only our lack of knowledge matters.

Let us label the six sides of the die with letters—A, B, C, D, E, and F—so as not to confuse the labels on the actual die (which are, of course, the numbers 1, 2, 3, 4, 5, and 6) with the number for the probability we are trying to compute. We want to compute the probability for a particular face, call this probability $p(E)$. From the assumptions that probabilities measure a degree of belief that a certain event will happen and that a greater degree of belief means that the probability is greater, we can derive several basic facts of probability.[12] First, all probabilities are real numbers between 0 and 1. A probability of 0 means that the event is certain not to occur. A probability of 1 means that the event is certain to occur. Second, if we have an exhaustive list of exclusive possible outcomes, then the probabilities of all these outcomes must add up to 1. If a list of outcomes is exhaustive, then by definition one or more of the outcomes is certain to occur. If we toss the die, at least one of the sides will come faceup, at least as far as we know. Remember, probabilities are about our knowledge, not about what will actually occur. It might be that the die will end up on one of its edges, but we've never seen a die that does this, so we assign this event a probability of 0. *Exclusive* means only one of the possibilities can be realized. We assume that only one face will be faceup after the toss. We will see A, or B, or C, or D, or E, or F. We will not see A and B or some other combination. Furthermore, if we see E, we also assume that this excludes seeing any other face in a single toss.

So in the case of the single die, we have six probabilities, one for each side. If we add these six numbers, we will get 1. Now the crucial step: we use the fact that we know of no reason to believe that one face is more likely to end faceup than any other. This means that we must assign the same probability to each face coming up. If we assigned different probabilities, it would mean that we really thought one face was more likely to come up than another, but by assumption, we think they are all equally likely. Since all the probabilities are equal, and they add up to 1, it follows that each individual probability must equal 1/6. In particular, we have $p(E) = \frac{1}{6}$.

Another way to think about this is to note that if we really think the six possibilities are equivalent, then changing the labels for the different events cannot change the probabilities. So if we interchange the labels A and B, so that A now means a 2 is faceup and B now means a 1 is faceup, the probabilities $p(A)$ and $p(B)$ are unchanged. Allowing similar relabelings among all the possible outcomes once again yields the fact that all six probabilities must be equal. But the relabeling emphasizes that it is our knowledge that has been expressed in the probability numbers.[13] Probability is definitely not something in reality.

We could in fact be wrong about the probability $p(E) = \frac{1}{6}$. If after throwing the die fifty times, we see the F come up every time, we reasonably begin to suspect (a long time before the fiftieth toss) that the die is loaded. Then our knowledge has changed: we now have the additional knowledge that tossing the die has yielded an F fifty times in a row, and there are standard procedures in probability theory to tell us how to modify the probabilities to take this new information into account. Our new assignment of probabilities might be wrong. It could be that in fact the die is honest and we have just had a very *improbable* run of luck to get fifty Fs in a row. By *improbable* I mean, of course, given our now assumed knowledge that the die is honest. From this assumption, we can compute the probability of getting fifty Fs in a row from an honest die is $(1/6)^{50}$, which is approximately 10^{-39}.

Let us suppose, however, that the die is in fact honest. Then it can be shown that, in the long run, we will see each side faceup in one-sixth of the tosses.[14] That is, it is *probable* we will see the side E faceup one-sixth of the time. We say that the *frequency* of side E is one-sixth. Once again, the word *probable* refers to our knowledge, in this case to the hypothesis that the E will appear one-sixth of the time.

To see how probability enters quantum mechanics, not as something "objective" in reality but as a consequence of our necessarily limited knowledge of this reality, let's go back to our model of a wave on the sea. As before, let us imagine that the wave is traveling from east to west, but now let us suppose that, some distance from the shore, the wave encounters a breakwater, with an opening 100 meters wide. That is, the wave is mostly stopped by the breakwater and can pass only through the opening. Except for the opening, the wave is completely stopped. Imagine that a large series of waves hits the breakwater and that the wave peaks are 10 meters apart. We say that the *wavelength* of the wave is 10 meters.

We want to understand the effect of this series of waves passing through the opening, so rather than the rectangular array of columns that we had earlier, let us imagine a single linear row, arrayed from south to north, to the west of the breakwater, the row being parallel to the waves before they hit the breakwater. Let us imagine that, rather than having paint on top of each column, we have a water-absorbing material, so that each time a wave overlaps a column, water is absorbed and the column gets taller. The height of the column measures how many times a wave has overlapped it. (We imagine that the water absorbed is somehow conveyed up the absorbent material, so the wave exerts its effect if it manages to overlap the original top of the column. The wave doesn't have to overlap the ever-increasing top of the column.) What will the columns in the single row look like after a large number of waves have passed through the opening in the breakwater?

Our intuition tells us—and our intuition is correct—that the columns closest to the breakwater have become higher than the others. The highest column is the one directly opposite the center of the opening. The columns immediately to the north and to the south of this column are also higher than they were before the waves began to pass through the breakwater, but not quite as high as the column directly opposite the breakwater. The columns still farther to the north or south will also be higher, but not quite as high as the columns immediately to the north or south of the central column, and so forth, until we come to columns to the north and to the south that have not been overlapped by waves at all and so are their original height.

Now comes a surprise. If we look still farther to the north or south, we will see columns that have been overlapped and so have gotten higher! This counterintuitive phenomenon is due to the constructive interference of the waves passing through the opening in the breakwater. A wave peak from one part of the opening has overlapped with another wave peak that has passed through the opening at a different time and a different part of the opening. In fact, if we were to look at the entire length of columns after a series of waves had passed through, we would see a finite series of higher columns, each with a central highest column, and separated by a column that kept its original height because it had never been overlapped.

This sequence of higher columns separated by a column that is unchanged in height is called an *interference pattern.* It is the hallmark of waves. But we will see this interference pattern only if the waves passing through the opening have the same wavelength and are parallel to the opening before they pass through. In such a case, we say that the waves are *coherent.* If the waves are not coherent, if the wavelength changes from moment to moment, for example, then the waves would still interfere constructively and destructively. We would just not be able to see this interference on the columns (or any other detection device).

If we now apply this model to the wave function of quantum mechanics, the height of the column would be the square of the amplitude of the wave function for an electron hitting that particular column. (Why it is the square of the wave function and not the wave function itself we will understand in a moment.) If a series of electrons were sent through the opening (we must imagine a much smaller opening for electrons, on the nanometer scale), then as the electrons hit, more would hit in regions where the heights were high, and fewer where the heights were small or unchanged. This is what we would see in a single universe. If we had a superhuman design, so that we could observe all the universes of the multiverse simultaneously, we would see electrons distributed continuously over an uncountable number of universes, hitting all columns simultaneously but more densely where the columns were higher. In the single universe that is all we can observe, we would say it is more probable that an electron will hit a column, whereas in the multiverse the electrons are more dense.

This will be clearer if we return to our wave model without the breakwater and with the rectangular array. Remember, Heisenberg showed that *if* a given column is first overlapped, then it is overwhelmingly likely that the single row of columns from west to east will also be overlapped. But Heisenberg could give no reason why one particular column in the first row was more likely than any other to be overlapped.

He could give no reason *because there is no reason.* In fact, in the multiverse, *all* the columns in the first row are overlapped! But we humans are machines designed so that if it is certain in one universe that only a single column is overlapped, we will with certainty see only that column overlapped. Linearity then forces us to see only a single column in the first row overlapped, even if they are all overlapped, as they are in this case. All the columns in the first row are overlapped, but we can see only one. We observers of the columns differentiate in the different universes of the multiverse: in one universe we see column A being overlapped, in another universe we see column B being overlapped, and so forth.

This is precisely analogous to the single-die example. There is no reason to prefer column A being seen overlapped rather than column B, or column C, and so forth, so we give each equal probability, for exactly the same reason that we gave each face of the die equal probability. But the probability is for which universe of the multiverse we will find ourselves in: will we find ourselves in the universe where column A is overlapped or the universe in which column B is overlapped? In the entire multiverse, all columns are overlapped, and in the entire multiverse, we see them all overlapped. But "we" in this sentence are the collection of our alternate selves in the alternate, but all equally real, universes of the multiverse.

This will be the case if the wave function weighs each column in the first row equally. In quantum mechanics, this equal weight is expressed by writing the wave function as a sum of terms, one term for each column, and each term multiplied by a number such that if the *squares* of all these numbers are added up, the result is 1. In the case of a uniform wave approaching N columns, the number multiplying each term would be $1/(N^{1/2})$. That is, it is the squares of the numbers that give the probability we will find ourselves in a universe in which column A is overlapped rather than any other column in the array of N first columns. If the numbers multiplying the terms are not equal, say the square of column A is twice the square of column B, then there are twice as many universes in the multiverse in which column A is overlapped as column B. Thus, before the wave reaches the first row of columns, we assign a probability of finding ourselves in a universe in which we see column A overlapped that is a factor of 2 larger than the probability we assign seeing B overlapped. In contrast to the die, we cannot improve the probability assignment with more observations, because quantum mechanics does not allow us to be aware of the other universes of the multiverse. We cannot, even in principle, get more knowledge. We have to be content with probabilities. Only a mentality that could see the entire multiverse would be able to dispense with probabilities. We shall discuss this further in Chapter 9, when we examine the Incarnation.

Before the measurement, we and our analogues in the different universes are identical in a way that the different sides of the die are not. If we were to look at the die faces under a microscope, we could probably find some differences among the faces. But quantum mechanics asserts something much stronger: systems in the same quantum states are identical in an absolutely fundamental way. If two systems in the same quantum state are interchanged, physical reality is unchanged. Absolutely no change at all. If we and one of our analogues were to be interchanged between two universes, quantum mechanics asserts that nothing physical would have occurred. At the most basic level, we and our analogues are completely identical. In other words, they are us in a most fundamental way. They are not merely our twins. They *are* us. They differ from us only in being in different localities in the multiverse. The best example of just how close these analogues are to us is the same person at different ages. Are you the same person you were ten years ago? In law and in common discourse, you are. If you had committed a serious crime ten years ago, you could still be liable for punishment. But you and your analogues in the multiverse are much closer in identity than you and yourself ten years ago. Your closeness is more like the identity of you and yourself a millionth of a second ago.

This quantum mechanical identity theory has far-reaching implications. It shows that the probabilities cannot be improved as long as we cannot be aware of the other versions of ourselves. But it also means that if any one of us were duplicated down to the quantum state, the duplicated copy would no longer be a copy. Instead, there would be two of the original person in this universe. There would not be an original and a copy. There would be two originals. Quantum mechanics says that there is no way, even in principle, of distinguishing between the original and the copy. There are two originals, but in different spatiotemporal locations in this universe. We shall consider this identity theory of quantum mechanics in Chapter 8, when we discuss the resurrection of the dead.

Why are the probabilities the squares of the weights rather than the weights themselves? The answer is that if the probabilities themselves were the weights, then destructive interference would never occur, since destructive interference requires some terms to be negative and probabilities must all be positive or 0. If destructive interference never occurred, then quantum mechanics would be no better than Hamilton-Jacobi theory, in which constructive interference unbalanced by destructive interference allowed concentrations of particles to build up to a singularity, which was destructive not just figuratively but really: physical reality itself was destroyed.

Another implication of the fact that we can be aware of only a single universe of the multiverse is the uncertainty principle, first derived by Werner Heisenberg. The uncertainty principle says that there is a limitation to the precision to which we can measure properties of a single particle. At the most basic level, this limitation is due to the fact that we cannot force a particle to belong entirely to our particular universe of the multiverse. More precisely, the uncertainty principle says that the product of the uncertainty in the position of a particle multiplied by the uncertainty in its momentum must always be greater than Planck's constant divided by 4π. (The *momentum* of a particle is the product of its mass and its velocity.) So the uncertainty principle doesn't say we can't be certain of anything. It merely says that there is a trade-off between uncertainties. Thus, we can be totally certain of the momentum if we are willing to be totally ignorant of the position.

Consider our model of a plane wave approaching the shore from the east. In this case, we know exactly what the momentum of the wave is: it is moving exactly from east to west. But we actually know nothing about the position of the complete wave. A true plane wave would have no end: it would stretch an infinite distance to the north and south, and there would be an unlimited number of peaks coming in from the east. We can obtain more knowledge of the wave's extent in the north-south direction if we imagine the wave hits a breakwater with an opening 100 meters in width. At the instant the wave hits the opening, we know its position north–south to within 100 meters—the width of the breakwater. But this knowledge comes at the price of losing our precise knowledge of its momentum.

Before the wave hit the breakwater, it was moving precisely from east to west. It had no component of velocity in the north–south direction. After having passed through the breakwater opening, the wave is spreading outward in all directions, even north and south. Watch a wave pass through a breakwater sometime, or perform the following experiment yourself. Place a cardboard barrier with a narrow opening in a pan of water. Set a plane wave in motion toward the barrier, say by moving a ruler held parallel to the barrier back and forth. You will see a wave passing through the narrow opening, a wave that is roughly circular and thus moving in part perpendicular to its original direction.

Now consider a group of surfers being carried by the wave through the opening in the breakwater. All will be moving mainly from east to west, but after passing through the breakwater, some will be moving a little toward the north, others will be moving a little toward the south. The area of the sea occupied by the surfers will get larger and larger as time advances (in quantum mechanics, this is called *wave packet spreading*). In the quantum mechanical situation, we cannot see all the surfers; we will be able to see only one. But which one will we see? That we cannot tell, even in principle. In one universe we will see a surfer going north, and in another we will see a surfer going south. Since all of our analogues in all the universes are identical, we cannot say which one "we" will see. Actually, "we" will see all the surfers. But in each universe a particular "we" will see only one surfer, with only one momentum. Before we interact with the group of surfers—and see only one—we can say only that the group of surfers has a spread, an uncertainty, of momentum. The uncertainty is due to the fact that there is a group of surfers but by our nature we can see only one. The ultimate reality is that the group of surfers has a group of momenta. We mistakenly think it has only one and attribute the spread of momenta to an uncertainty in nature herself. There is no uncertainty in nature. Quantum mechanics is deterministic, and in fact, as I showed earlier, quantum mechanics is classical mechanics modified to make the determinism certain. The uncertainty in quantum mechanics, like the probability in quantum mechanics, is a manifestation of human ignorance, our ignorance of, and inability to communicate with, the analogues of ourselves in the other universes of the multiverse.

There is also no *quantum nonlocality,* a term that refers to a spooky action in which it is claimed that a distant object "knows" instantly what has just happened in a laboratory. In fact, as we shall see shortly, knowledge cannot be propagated faster than light. An object a light-year distant cannot "know" what has happened in a laboratory on Earth in a time less than a year. The appearance of nonlocality is due to ignorance of our analogues in the multiverse, all of whom are experimenting on a coherent quantum state.

Quantum nonlocality is usually discussed in the context of the Einstein-Podolsky-Rosen (EPR) experiment, or the Aspect experiment, but the basic idea is more easily understood in terms of a system of two electrons formed in such a way that the spins of the electrons are always in the opposite directions. An electron is always spinning like a little top, but in contrast to the tops that we played with as children, an electron can have only two possible spin directions. The Earth is also spinning on its axis like a top, and the spin direction is along the top's axis of rotation. If we curl the four fingers of our right hand in the direction of the top's rotation, our thumb will by definition point in the direction of the spin. The Earth's spin could point in principle in any direction, and in fact its spin direction is changing slightly all the time. If we measure the component of the Earth's spin perpendicular to the plane of its orbit, we will find that only part of its spin is perpendicular to the plane of its orbit. The Earth's spin axis is tilted with respect to the plane of its orbit, and this is what causes the seasons. If we try to measure the spin of an electron, however, we will get one of only two answers. If we try to measure the component of the spin of an electron in the up-down direction, we will get the answer that either all of the electron's spin is oriented in the up direction or all of its spin is oriented in the down direction. We will never observe only a fraction of the spin oriented in the up-down direction. With the electron, in contrast to the Earth, it is all or nothing.

The same would be true if we had decided to measure the electron spin in an east-west direction. Our instrument would tell us that either all of the electron's spin pointed toward the east or it all pointed toward the west. The instrument would never tell us that only some of the electron's spin pointed toward the east or west. East or west, or up or down, the electron spin is always pointing in one definite direction, determined by the set of directions we have decided to measure. This suggests that we and our analogues are playing a big role in determining what spin the electron is actually observed to have, and that is correct.

Now let us analyze what we would see if we tried to measure the spins of two electrons that are formed in a coherent state, the state in which the spin of one electron is always opposite the spin of the other. If we decided to measure both spins in the up-down direction, we would measure one electron to have spin up and the other to have spin down. Apparent nonlocality arises because this would have to be true even if, before the measurement, we allowed the two electrons to travel apart, so that one electron remained in the laboratory while the other traveled a light-year away. Measuring the spin of the electron in the laboratory to be spin down would tell us that the spin of the faraway electron would have to be up. In fact, this is exactly what experiments show. If two far-separated observers measure the spins of the two electrons in this special coherent state, one observer will measure his electron to have spin up, and the other will measure her electron to have spin down. This will be the observation no matter how close together in time the two observers make their measurements. The two measurements can be set up so there is no time for a signal, moving at the speed of light or slower, to travel from one laboratory to the other. So how does the second electron to have its spin measured know the result of the other experiment? It seems that the electrons are using some sort of faster-than-light communication.

Not true. What is happening is that the analogues of each of the two observers are differentiating in the universes of the multiverse. However, the two electrons are in a coherent state, and this coherence is respected across the multiverse. When the analogues of the observer in the first lab measure their electrons in the universes of the multiverse, one set of analogues will measure the electron to have spin up and the other analogues will measure the electron to have spin down. When the analogues of the observer in the second lab measure their electrons, once again one set will measure the electron to have spin up and the other set will measure it to have spin down. Since the electron by its very nature cannot yield any other result, this should not be surprising.

But the linearity of quantum mechanics and the coherence of the electron state force a correlation between the measurements in the universes. Think about it. If an electron was created in the spin-up state, the device used to measure the spin had better measure it to be spin up. Otherwise, we junk the device and buy a new measuring instrument. In the coherent state of the two electrons just described, the electron is in a superposition of the two actualities—spin up in half the universes, and spin down in the other half—so in half the universes, the first observer sees the electron to have spin up, and in the other half of the universes, he sees it to have spin down. Ditto for the second observer. Since both observers have correctly functioning instruments, if one observer measures the electron to have spin up, the other must measure it to be spin down. And this is what happens.

This would not be a surprise if one electron was set beforehand to be definitely spin up and the other definitely spin down. You set the electron to have a definite spin, and you observe the electron to have the spin you set up. The coherence and linearity force the universes to differentiate coherently: in the universes in which the first observer set measures his electron to have spin up, the second observer set measures her electron to have spin down. The reverse is also true. In the universes in which the first observer set measures his electron to have spin down, the second observer set measures her electron to have spin up. Linearity of quantum mechanics forces this result because the spins are correlated by construction. They have to be opposite in each pair of universes, and they are. The correlation is carried with the electrons at the speed they are moving apart, necessarily slower than light.

If the two observers had decided instead to measure the spins of the electrons in the east-west direction rather than up-down, one observer set of the first electron would measure his electron to have spin east, and the other observer set of the first electron would measure her electron to have spin west. There would be two corresponding observer sets for the second electron. One observer set would measure the second electron to have spin east, and the other observer set would measure the electron to have spin west. But again the observer sets of the two electrons would be correlated. The observer set of the first electron who measured his electron to have spin east would be in the same set of universes as the second observer, who measured her electron to have spin west. And the reverse correlation would also hold.

It is often forgotten that there is a third measurement in these EPR nonlocality experiments: the measurement that compares the results of the two experiments. Since experimenters carry out this measurement by signaling to each other, the measurement is the result of exchanging information at a speed less than that of light. This third measurement, by linearity, carries along the original coherence, or correlation setup, so the experimenters always see that, whatever spins they decide to measure, they observe opposite spins if in fact they have set up their experiments to measure spins along the same axis. If they decide to measure the spins along different axes, then the analysis is a bit more complicated—the differentiation of the universes of the multiverse is more involved—but once again it is easy to see that there is no nonlocality. Anyone who claims otherwise does not accept the existence of the multiverse, which is to say that he or she does not accept quantum mechanics.

The final manifestation of the multiverse that we shall consider is the phenomenon of *quantum tunneling.* Return once again to our example of a wave moving from east to west and encountering a breakwater. Surfers are traveling on the crest of the wave. This time, however, suppose there is no opening in the breakwater, and the wall making up the breakwater is very high, 10 meters higher than the wave crest height. It looks as if all the surfers are doomed. Most are indeed doomed: most of the surfers and their waves bounce off the formidable wall and are carried back out to sea. But a tiny fraction of the surfers (and their waves) are later to be found on the west side of the wall, moving with diminished speed from east to west. They are said to have quantum mechanically "tunneled" through that wall.

How did they actually get through? They didn't drill a hole through the wall, as the name of this phenomenon suggests. What actually happened was that the wave incident on the wall was not infinite in its west-east extent. This means that the actual wave consists of waves of varying heights, *some of which are higher than the wall.* It is only the average height of the waves that is lower than the height of the wall. The surfers on the crests of these higher-than-the-wall waves, not surprisingly, manage to travel over the wall and later are to be found on the other side.

In a more realistic quantum mechanical example, an electron is traveling toward a *potential barrier,* which is a region where the voltage is so great that the kinetic energy (the energy of motion) of the electron is insufficient to let it pass through. If there were only one universe, the electron would be repelled by the potential barrier (high voltage) and never pass through. However, the uncertainty principle tells us that if we know roughly when the electron hits the barrier, then the electron's momentum in the direction of the barrier must be "uncertain," which in turn just means that some of the electron's analogues in the multiverse have a variety of momenta, and since the kinetic energy is proportional to the square of the momentum, some of these analogues will have kinetic energy greater than the barrier. These analogues will, not surprisingly, penetrate the barrier.[15] As in our wave example, the average energy of the electrons throughout the multiverse is insufficient to surmount the barrier, but some of the electrons in the multiverse have energy higher than the average. It is also in principle possible for a particle in a definite universe to "borrow" for a very short time (the allowed time being given by the uncertainty principle)[16] an arbitrarily large amount of energy from their analogues in the multiverse. I shall use this fact in Chapter 8 in my explanation of the Resurrection of Jesus.

The Theory of General Relativity

With quantum mechanics, relativity forms the fundamental foundation of modern physics. The name of the theory is a huge misnomer; Albert Einstein himself initially called his theory "invariant theory" or "theory of absolutes." Einstein's terminology is more accurate. The theory of relativity is indeed built upon an absolute: the speed of light in a vacuum. Relativity asserts that all observers, whatever their velocities relative to one another, measure exactly the same number for the speed of light. The invariance of the speed of light allows two apparently unrelated quantities, the dimension of time and the three dimensions of space, to be united into one four-dimensional entity, spacetime.

Spacetime can be visualized in a simple picture, the Minkowski diagram, illustrated in Figure 2.3.

The key idea is to use the same units for both the time and the space directions. We routinely use the same units in all of the three spatial dimensions. We don't, for example, use yards when we measure distances in the north-south direction and meters when we measure distances in the east-west direction. Using the same units in both cases allows us to combine the two measurements to get a net distance traveled when we go 4 meters due north followed by 3 meters due east. The Pythagorean theorem then tells us that we are 5 meters from our starting point.

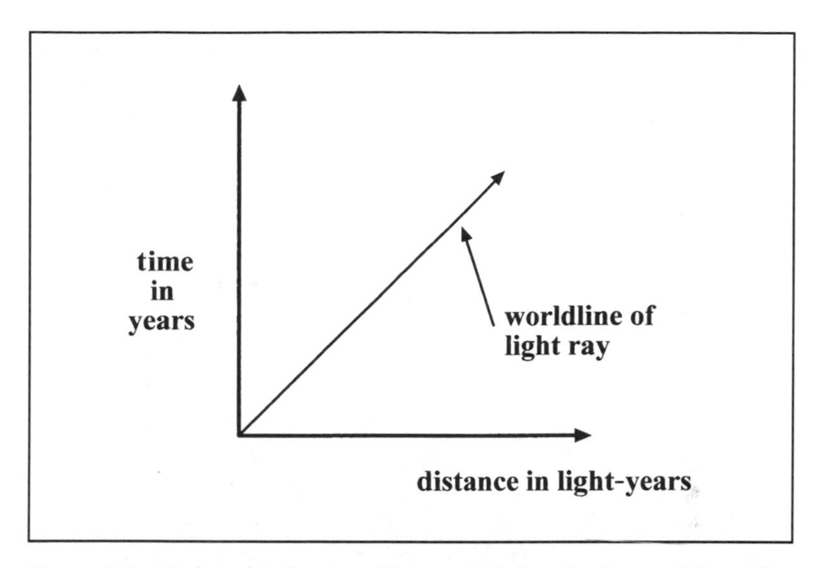

Figure 2.3. Minkowski diagram. The vertical line is the worldline of a particle that never moves, and the horizontal line is the spatial universe at time t = 0. The units are chosen so that light moves at one unit of space for one unit of time. One light-year is the distance that light travels in a year. This means that light has a worldline at a 45-degree angle.

We do the same thing when we measure times, distances, or a mixture of the two in spacetime. We measure times in years, and distances in light-years. Notice that the speed of light has come in, because in defining the standard distance *light-year,* we use the speed of light in a vacuum. Why is light so central? Because the particles of light, the photons, have zero mass. Any particle that has zero mass will move at this central speed. It just so happens that the photon was the first particle to be discovered that had zero mass. As we shall see in the next section, however, all fundamental particles have zero mass. A zero-mass particle necessarily moves at the speed of light.

In these natural units for spacetime, the speed of light is 1: a photon moves at 1 light-year per year. In Figure 2.3, the path of a light ray is pictured. It makes an angle of 45 degrees with respect to the vertical axis, which is the time axis (time increases as you move up the page), and also 45 degrees with respect to the horizontal axis, which is the spatial axis (for convenience of visualization, two spatial dimensions have been omitted). All objects with nonzero mass have to move at a speed less than the speed of light, which means that, if we were to draw the history of such an object in a Minkowski diagram, it would be shown to travel more in the time direction than in the spatial direction in comparison to a light ray. An object that doesn't move at all, for example, has a path—a *worldline,* to use the technical term—that only moves up the time axis: it has no spatial component at all.

If we rotate the path of the light ray pictured in Figure 2.3 around the time axis, in effect adding another spatial dimension, a cone will be formed. This is called, appropriately, the *forward (or future) light cone,* and it is of crucial importance, because all particles starting out from the origin of spacetime coordinates must necessarily move inside it. This is just another way of saying that nothing can go faster than light. The light cones at all points (the technical term for a "point" of spacetime is *event*) define the *causal structure* of spacetime, because if you want to send a signal to someone at another event (a different spatial place at another time), then the signal worldline must lie inside the forward light cones at every event. Conversely, something can affect you at your spatial location at a definite time only if it occurred in your past (backward) light cone. If an event lies outside your past light cone, it can have no effect on you, since no signal and no effect from that event can reach you. It could reach you only by traveling faster than light, and this is impossible.

Why can one not signal or travel faster than light? There are several answers to this question that can be found in the elementary physics textbooks. For example, it can be shown that an infinite amount of energy would be required to accelerate a particle with nonzero mass to the speed of light. But these observations merely show that relativity theory is consistent. What is wrong with the Newtonian theory of mechanics, which allowed one to reach arbitrary speeds?

The central difficulty with having no ultimate speed limit is that it allows a breakdown in determinism, somewhat analogous to the singularities in the laboratory whose avoidance forced us to accept quantum mechanics. In the absence of an ultimate speed limit, there is no limit to the energy that can be extracted from an object by letting it undergo gravitational collapse. In relativity theory, there is such a limit, because if the object got small enough, it would form a black hole, which would prevent any further energy extraction. If energy extraction is unlimited, it can be shown that it would be possible to use this unlimited energy source to propel an object to spatial infinity in finite time. What happens then? We cannot say, because this infinite energy extraction in the laboratory has generated a singularity. To prevent this singularity, to preserve determinism, an ultimate speed limit must exist.

Determinism in relativistic quantum mechanics is called *unitarity.* Determinism in all forms of physics is concerned with time evolution. Since in quantum mechanics the basic, most fundamental entity is not the particle but instead the wave function that is in one-to-one correspondence with a particle and its analogues in the multiverse, it is the time evolution of the wave function that is subject to determinism. The time evolution of the wave function is controlled by the *time evolution operator* $U(t, t_i)$, which carries the wave function at some initial time t_i *uniquely* into a wave function at some later time t. This is expressed in very simple equation form as $\psi(t) = U(t, t_i)\psi(t_i)$, where $\psi(t_i)$ is the wave function at the initial time t_i, and $\psi(t)$ is the wave function at some later time t. Determinism is expressed by the crucial word *uniquely.* That is, whatever the initial wave function is, there will be one and only one later wave function generated by the time evolution operator from this initial wave function.

The inverse time evolution operator $U^{-1}(t, t_i)$, where the exponent (-1) just means "inverse," undoes the effect of the original time evolution operator. That is, the inverse time evolution operator acting on the later wave function $\psi(t)$ carries us back in time to the earlier wave function $\psi(t_i)$. This also has a very simple equation expression: $U^{-1}(t, t_i)\psi(t) = \psi(t_i)$. Now for the key point. The mathematical consequence of two assumptions—first, that the time evolution operator acts on *all* possible initial wave functions, and second, that the later wave function obtained from any of these possible wave functions is unique (that is, determinism holds)—is that the inverse time evolution operator $U^{-1}(t, t_i)$ both exists and is related in a very simple way to the time evolution operator.[17] Specifically, the inverse time evolution operator must be what is called the *Hermitian conjugate* of the time operator. This means if we were to write the time evolution operator as a matrix of numbers, the standard square array of numbers, the inverse is obtained by interchanging rows and columns in the original matrix while replacing each number with its complex conjugate. An operator whose inverse is obtained from the original operator in this simple way is called a *unitary* operator. The requirement that the time evolution operator be unitary is called *unitarity.*

If you did not follow the full meaning of the mathematical terminology in the preceding paragraph, don't worry about it. The details are not important. What is most important is the fact that unitarity is an expression of determinism in relativistic quantum mechanics. It is also important to note that unitarity is not quite the same as, but is intimately related to, the law of conservation of energy. Physicists have constructed models of unitarity violation to see how much is experimentally allowed. Even a tiny amount of nonunitary time evolution would be disastrous: if you were to turn on your microwave oven, so much energy would be created out of nothing that the Earth would be blown apart! I shall assume that unitarity holds.

The determinism implied by unitarity is a very strong sort of determinism. Since the inverse time evolution operator exists, we can think of determinism as working backward as well as forward in time. In most discussions of determinism, it is assumed that the future and the present are determined by what happened in the past. But unitarity tells us it is equally correct to think of the future state of the multiverse as determining the past and present states. In philosophy, this future determinism is called *teleology* and is thought to be unscientific. Not so: teleology is alive and well in physics. But in physics, teleology is called "unitarity." What is happening in the universe and multiverse today is determined by the goal in the far future that has been set for the universe and the multiverse.

The consequences of unitarity are vast, and we shall discuss them throughout this book. One is the Bekenstein Bound on the information content of any physical system. The amount of information inside a sphere of radius R, with the system inside the sphere having no more than the mass-energy M, is

$$\text{Information} \leq (2.6 \times 10^{43} \text{ bits}) \times (M/[1 \text{ kg}]) \times (R/[1 \text{ m}])$$

The binary *bit* is the most basic unit of information. These days, however, the size of computer memory is usually given in *bytes*—1 byte is equal to 8 bits. So divide the coefficient by 8 to get the ultimate limitation on the information content, which is also the ultimate limit on its complexity, on *any* physical system. I pointed out in my earlier book, *The Physics of Immortality,* that the Bekenstein Bound can also be thought of as an expression of the relativistic version of the uncertainty principle.

A human being is a physical system, and so is subject to the Bekenstein Bound. Most people weigh less than 100 kilograms, and almost everybody can bend over so as to fit, barely, in a sphere of radius 1 meter. Therefore, the complexity of a human being is less than 3.2×10^{44} bytes. A typical hard-drive memory capacity these days is 30 gigabytes, so any human being can be coded using a mere 10^{34}, or 10 billion trillion trillion, such hard drives. However, it should be kept it mind that this number is an upper bound. The Bekenstein Bound guarantees that a human being can be coded down to the quantum state, that is to say exactly, using this amount of information. In actual practice, what is essential to human identity can be coded with much less information. The Bekenstein Bound number codes not only the essentials of your personality but also the exact location of every one of your hairs. If one of your hairs were to be moved a trillionth of an inch, the Bekenstein Bound information would change, but the essential you would not.

The Bekenstein Bound also constrains how much complexity there can be in the molecular disorder in the universe. The molecular disorder of a physical system is quantified by its *entropy,* and methods to compute the entropy of any system can be found in any textbook on thermodynamics. The Second Law of Thermodynamics governs molecular disorder. This law asserts that the entropy of the universe must never decrease. Thinking or feeling on the part of any living being requires the increase of entropy, so the Second Law of Thermodynamics can be regarded as a law governing the spiritual side of the material universe. The fundamental unit of thinking, which is a form of information processing, is 1 byte, which equals 7.655994×10^{44} J/°C, where Joules per degree Celsius is the physical unit of entropy. Not only is there a physical law that governs thinking and feeling, but we physicists know the conversion factors to high precision!

One of Einstein's great achievements was extending to spacetime the idea that gravity is curvature. Even in Newtonian theory, gravity is curvature. This is not what is usually taught in the elementary textbooks, which say that before Einstein, gravity was pictured as a force, but it is true nonetheless. The great French mathematician Elie Cartan (1869–1951) showed in 1923 that the gravitational field of Newton is actually not a force either but instead curvature in time only.[18] In Newtonian gravity, space was not curved, only time. Since the real underlying reality is not space and time separately but space and time unified into spacetime, one would expect that it would be spacetime, and not just time, that is curved. Einstein found a theory of curved spacetime, the Einstein gravitational field equations.

There is one difficulty with Einstein's theory of gravity, namely that its Hamilton-Jacobi version generates singularities in the laboratory. That is, it is a classical theory that needs to be "quantized"; to put it another way, it needs to be made fully consistent with the linear superposition principle, the principle that prevents these nasty, nonexistent singularities in the laboratory. The quantization of Einstein's gravity theory was actually achieved in the 1960s by two American Nobel Prize–winning physicists, Richard P. Feynman and Steven Weinberg, who, remarkably, did not realize they had solved the problem of quantum gravity. (Most physicists don't realize this even today.) They were expecting to obtain a quantum theory of gravity that had derivatives no higher than second order. Unfortunately, general relativity, the principle that the laws of physics are observer-independent, will not be completely consistent with the linear superposition principle unless derivatives of an arbitrary higher order are present. Feynman and Weinberg discovered this and then wrote down the essentially unique quantum theory of gravity that followed from gravity being spacetime curvature consistent with the linear superposition principle. But they recoiled in horror from the theory they had discovered, even though they knew that it both was mathematically consistent and agreed with all experiments conducted to date. Furthermore, they realized that it was the unique consistent quantum general relativity theory that could be derived from the idea that gravity is spacetime curvature, an idea that is the only natural generalization of the Newtonian theory of gravity as temporal curvature.

Feynman, Weinberg, and most subsequent physicists have not accepted this unique theory of gravity because they could not accept its philosophical implications. All previous theories of physics have been built on equations called *partial differential equations,* which basically equate derivatives of various physical quantities. In the past, the fundamental equations have had no higher than second order derivatives, meaning that there were only a finite number of terms. We might not have been able to determine the initial conditions to feed into these equations with sufficient precision to predict the future—remember the uncertainty principle—but at least we could determine the equations themselves with certainty.

What Feynman and Weinberg really discovered was another and more fundamental limitation on human knowledge: not only can we not, even in principle, determine exactly the position and momentum of a particle, but we cannot even determine or write down, even in principle, the ultimate equations the particle will follow! In fact, if we consider an equation to consist of an infinite number of terms, there *are* no ultimate equations! This does not mean that the history of the particle and its analogues in the multiverse is not subject to, and completely determined by, physical law. It is, even in the Feynman-Weinberg theory. But we humans will never know this theory. As we shall see in the next chapter, where I will develop the full implications of this theory, our descendants in the far future will be able to understand this theory with ever-increasing precision, but not even they will completely understand it until the very end of time.

The Standard Model of Particle Physics

Relativity theory accounts for one "force," the force of gravitation. All other observed forces in nature are described correctly, as far as all experiments carried out to date have been able to tell, by the Standard Model, first developed in the 1960s. According to the Standard Model, there are two basic types of fundamental particles. The first type make up the building blocks of matter and are called *fermions* because, like the first fundamental fermion to be discovered, the electron, they have two possible spin directions, and Enrico Fermi (along with Paul Dirac, who didn't get name credit, alas) was the first to understand the implications of such particles for thermodynamics. These fundamental fermions are usually called spin one-half particles, because the value of their spin angular momentum is one-half of the fundamental constant of quantum mechanics, Planck's constant divided by 2π. (Planck's constant divided by 2π is the more basic form of the constant and is called the reduced Planck's constant.)

The other type of fundamental particle carries the forces between the fermions; these particles are called *bosons,* named after the Indian physicist Sir Jagadis Bose, who, along with Albert Einstein, was the first to understand the implications of these sorts of particles for thermodynamics. The bosons all have angular momentum that is an integer multiple of the reduced Planck's constant. Photons, the carriers of the electromagnetic force, are bosons with spin 1. All the fundamental force-carrying bosons of the Standard Model have spin 1. Gravity is carried by a boson called the *graviton,* a boson with spin 2. There is one other boson in the Standard Model, the *Higgs boson,* which has spin 0. The Higgs boson is a unique particle, neither a building block particle nor a force-carrying particle. We shall discuss its role in nature a bit later.

The fundamental fermions of the Standard Model are further subdivided into *leptons* and *quarks,* and each of these comes in three families. The families of leptons are arranged as in table 2.1.

Table 2.1

Leptons

First Family	Second Family	Third Family
ν_e	ν_μ	ν_τ
$e-$	$\mu-$	$\tau-$

The symbol e− represents the electron, and the superscript means it has a negative charge. The symbol μ− represents a particle called the *muon,* and it has the same electric charge as the electron, in both magnitude and sign (negative charge). The muon is about 200 times as massive as the electron. It is usual to give the masses of the fundamental particles in energy units (remember $E = mc^2$), and the energy unit used is the electron volt, eV, the energy that would be gained by an electron passing through a potential of 1 volt. One million electron volts is written 1 MeV. The electron's mass is 0.51 MeV, and the muon's is 106 MeV. The muon is the second family's version of an electron. The third family's version is the *tauon* (or tau particle), represented by the symbol τ−. Its mass is 1,784 MeV. Each of these "flavors" of charged lepton has an associated *neutrino,* represented by the symbol ν with the subscript of the charged lepton with which it is associated. The leptons in each family should be considered as two distinct states of the same particle. The electron is a charged particle, and the electron neutrino is the same particle with its charge (and almost all of its mass) given up. The neutrinos are known to have nonzero mass, or rather at least two of the three neutrinos do, but the masses are very small and have not yet been measured. The quarks also come in three families, arranged as in table 2.2.

Table 2.2

Quarks

First Family	Second Family	Third Family
u	*c*	*t*
d	*s*	*b*

The quarks are the building blocks of the protons and neutrons, the particles that make up the nuclei of atoms. The quarks in the first row, the *u, c,* and *t* quarks, called the up quark, the charmed quark, and the top quark, respectively, all have an electric charge equal to +⅔ the magnitude of the charge on the electron. The quarks in bottom row, the *d, s,* and *b* quarks, called the down quark, the strange quark, and the bottom quark, all have an electric charge of −⅓ the magnitude of the charge on the electron. Quarks are never seen as separate particles; they are always bound together in groups of three. This makes it difficult to determine the masses of the lightest quarks, the up quark and the down quark. The indirectly determined mass of the up quark is 5 MeV, and the down quark mass is 9 MeV. (I should mention that I am in a minority group of physicists who believe that the up quark actually has zero mass, because such a mass would resolve a problem with the Standard Model called the strong CP problem without having to invent physics beyond the Standard Model. Eventually, experiment will decide who is correct. In this book, I shall, of course, assume that I am.) The strange quark has a mass of about 175 MeV, the charmed quark mass is 1,270 MeV, the bottom quark, 4,400 MeV, and the monster top quark, 175,000 MeV = 175 GeV.

Only the two lowest mass quarks make up familiar nuclear particles. The proton is two up quarks and one down quark, which we write as $p = uud$. The neutron is one up quark and two down quarks, $n = udd$. As with leptons, the two quarks in each family should be thought of as two different states of the same particle. So the decay of the neutron into a proton (about half of a batch of free neutrons will decay into protons in about thirteen minutes) is actually a transition of one of the down quarks into its less massive state, the up quark.

These leptons and quarks are the true atoms of nature. Remember that the word *atom* in Greek means "not capable of being cut into smaller pieces." The "atoms" of chemistry, such as carbon and oxygen, are not atoms in the Greek sense, since they can be subdivided into nuclei and electrons, and the nuclei can be subdivided into neutrons and protons, and the neutrons and protons can be subdivided into quarks. But as far as we can tell, the quarks and the electrons cannot be further subdivided. And we have tried hard to subdivide them. We have slammed electrons together at higher and higher energies, and they still behave as if they were indivisible point particles. All nonatoms we have ever seen have come apart long before they were smashed with energies equal to their mass. (To get a feel for what this means, a human's mass corresponds to the energy released by a 1,000-megaton nuclear bomb. This is about 100,000 times the energy released by the atomic bomb that annihilated the city of Hiroshima.) We have hit an electron with energies greater than 50 GeV, which is 250,000 times its mass. No substructure to an electron has ever been seen.

Each of the fundamental leptons and quarks has a corresponding *antiparticle.* The antiparticle of any particle has the same mass as the particle, and the same spin. But the antiparticle has the opposite charge. For example, the antiparticle of the electron has the same mass as the electron and has a spin of one-half. But it has a positive charge, is appropriately called a *positron,* and is represented by the symbol e^+, or $\bar{e}$. In general, the antiparticle of any particle is represented by the same symbol as the particle but overlined. Composite antiparticles are made up of antiquarks and antileptons. For example, an antiproton is made up of three antiquarks, specifically $\bar{u}\,\bar{u}\,\bar{d}$. Since the antiparticles have charges opposite those of the corresponding particles, the charge of the antiproton is (in units of the magnitude of the charge on the electron) $(-2/3) + (-2/3) + (+1/3) = -1$; that is, the same charge as the electron. The antiproton has the same mass as the proton, but it has the opposite charge.

All of the leptons in the lepton chart are said to have 1 unit of *lepton number,* and each family has a different lepton number. Each of the three lepton numbers is separately conserved in every particle interaction that has been seen to date (there is a single exception, called *neutrino mixing,* which will not concern us). The antileptons have the opposite lepton number. Thus, an electron has an electron lepton number +1, and a positron, or antielectron, an electron lepton number of −1. The electron neutrino has an electron lepton number of +1, and the electron antineutrino, −1.

All the quarks in the quark chart have +⅓ units of *baryon number.* (*Baryon* means "heavy," and the quarks make up the particles, protons and neutrons, that are heavy relative to the mass of the electrons. The word *lepton* means "small" in Greek.) Like lepton number, baryon number is conserved, but there is no separate baryon conservation law for each family.

Antiparticles collectively are called *antimatter* because if corresponding antiparticles are brought together, they mutually annihilate, giving energy. For example, if an electron and a positron are brought together, they annihilate into two gamma rays, which are high-energy photons. Notice that in this annihilation, electric charge is conserved, since the charges on the two particles are equal and opposite, and lepton number is conserved, since the lepton numbers of the two particles are also equal and opposite.

There are three fundamental boson force fields, the U(1) field, the SU(2) field, and the SU(3) field. The SU(3) field has a name, the *color force.* The color force is what binds the quarks together to form the nucleons, the neutron and the proton. The color force is carried by a particle called the *gluon,* so named because it glues the quarks together. The gluon has eight distinct states, distinguished by different color "charges," which are unrelated to, but the same nature as, the electric charge that is carried by an electron. The gluon is thus different from the photon, which carries the electromagnetic force but does not have an electric charge. The fact that the gluons both carry the color force and have a color charge has an important implication: the magnitude of the color force increases linearly with distance, unlike the electric force, which decreases as the inverse square of the distance. All particles that carry a color charge—all quarks and, of course, gluons—are very tightly bound together. Thus, the Standard Model tells us that we will never see individual quarks or gluons. We will see only certain combinations of quarks for which the color charges can be shown to cancel out. Such particles are called *colorless,* and the familiar examples are the proton and the neutron. For these particles, there is still a residual color force, and it is called the *strong force,* the force that binds neutrons and protons together in the atomic nucleus.

The other two fundamental fields of the Standard Model, the U(1) and the SU(2) fields, together generate the *electroweak force.* As the name suggests, according to the Standard Model, the electromagnetic force is not fundamental, nor is the weak force, which is responsible for phenomena such as radioactive decay. We have already mentioned the decay of the neutron into a proton. This is an example of a decay caused by the weak force. The electromagnetic force and the weak force are different manifestations of the electroweak force, just as the electric force and the magnetic force are different manifestations of the electromagnetic force.

The electromagnetic force and the weak force are formed from the U(1) and SU(2) fields by their interaction with another field, the *Higgs boson field.* According to the Standard Model, the Higgs field is an enormously powerful force field that permeates all of space, with a constant density throughout the universe of about -10^{26} grams per cubic centimeter (gm/cm^3). This density of the Higgs energy field is to be compared with the density of water, 1 gm/cm^3, or the density of air, about 10^{-3} gm/cm^3. In other words, the Higgs field is one hundred trillion trillion times as dense as water. Furthermore, the density of the Higgs field is not positive, as are all familiar material fields, but negative.

The Higgs field causes the U(1) and SU(2) fields to rearrange themselves, resulting in the electromagnetic and weak forces that we observe in nature. But the Higgs field does more: it causes the particles that carry the weak force to have a mass. There are two particles responsible for the weak force: the W^+ and the Z^0, where the superscripts denote the electric charges of the two particles. The W^+ spin-1 boson carries 1 unit of positive charge (so its antiparticle, the W^-, has a negative charge), while the Z^0 spin-1 boson has 0 electric charge. Neither particle interacts with the color force.

The masses that the *W* and *Z* particles pick up from the Higgs field, which permeates the universe, are enormous: the *W* has a mass of 80 GeV, and the *Z* has a mass of 91 GeV. By comparison, the proton and neutron both have a mass of about 1 GeV. So the *W* and *Z* particles have a mass nearly as great as that of a silver atom (104 GeV). Relativistic quantum mechanics tells us that the range and strength of a force depend on the mass of the particle that carries the force, with the effective distance and strength decreasing as the mass increases. The large masses of the *W* and *Z* particles are the reason the weak force is weak.

The *W* boson and its antiparticle are responsible for all weak interaction decays that involve a change in electric charge. For example, they are responsible for the decay of the neutron into a proton. Recall that a proton is made up of three quarks *uud,* while the neutron is made of the *udd* triad. In neutron decay, a down quark with charge $-1/3$ undergoes a transition to an up quark with charge $+2/3$. It makes this transition by emitting a W^- boson, with charge -1, so electric charge is conserved. But the *W* boson is very massive, and it is itself unstable. It quickly decays into an electron and an electron antineutrino. In this decay, notice that electric charge is conserved (the electron has the *W* boson's electric charge), as is lepton number. The electron has lepton number +1, and the electron antineutrino has lepton number -1, and the *W* is not a lepton, so it has lepton number 0: $0 = (+1) + (-1)$. The weak decay can therefore be written as $d \rightarrow u + W^- \rightarrow u + e^- + \bar{\nu}_e$ or because of the color force, which confines the quarks to remain inside baryons of three quarks, we can write this decay as $n \rightarrow p + e^- + \bar{\nu}_e$. The *W* decays so fast that we never see it.

The *Z* boson is responsible for the scattering of particles where there is no electric charge interchange. A neutrino can scatter an electron, for example, by exchange of a *Z* boson between the two particles.

The Higgs field gives a nonzero mass to the *W* and *Z* bosons. Remarkably, the Higgs field gives masses to *all* the particles. All leptons and all quarks, in other words, are massless in the absence of the Higgs field. As the leptons and quarks move through the Higgs field, which fills the universe, they interact with it, and this interaction acts as a resistance to a change in velocity (this is what mass does), somewhat analogous to the way a viscous fluid such as syrup resists the motion of a particle through it. The strength of the interaction determines the mass. The more massive a lepton or quark is, the stronger its interaction with the Higgs field. The neutrinos (and possibly the up quark) have the least interaction, and the top quark has the most.

Since all observed atoms are made up of first family leptons and quarks, one might wonder, what is the point of having three families? Why hasn't nature been more parsimonious and made do with only one family? There may be an answer in the fact that the universe contains more matter than antimatter. In the very early universe, when the density of radiation throughout the universe was much higher than it is today, particle-antiparticle pairs, such as electrons and positrons, were continuously being formed out of the radiation field. Thus, one would expect that the universe would have an equal amount of matter and antimatter. But the universe does not, and a good thing too. If the amounts of matter and antimatter had been equal in the very early universe, these equal amounts would with high probability have been completely annihilated by now, leaving no matter around to form stars, planets, and human beings. We exist only because there is a net amount of matter over antimatter, and there has been this imbalance ever since very early in universal history. Why?

The Standard Model provides a process that will create more matter than antimatter out of the early universe radiation field, but remarkably, this process works *only* if there are at least three families of leptons. So to have human life around, we need at least three families of leptons and quarks, and this minimum number is exactly what we have. The Standard Model process uses the SU(2) field to create what might be called "kinks" in the "vacuum." "Vacuum" is put in quotation marks because this vacuum is not without structure, as the word *vacuum* usually implies. Think of this vacuum as related to the Higgs energy field, where, since it is universal, we can rescale our measure of the 0 of energy in the universe so that the Higgs energy field is the official 0.

Every time one kink is formed in the vacuum, one net lepton or quark is created. This process violates the lepton and baryon conservation laws and has never been seen. The Standard Model gives a very good reason why we have never observed this process: it takes an enormous amount of energy to create this kink in the vacuum. A collision of an electron with a positron in which each particle had 1 TeV, or 10,000 GeV, of energy would do it, but our most powerful electron-positron colliders have managed to reach only about 50 GeV per particle. This process that can create matter without creating an equal amount of antimatter might be responsible for the net amount of matter in the universe. If it is, then the vacuum of the universe has some kinks in it. The Standard Model assures us that this process exists, and the Standard Model has been confirmed by all tests performed to date. So the process probably exists. Conversely, if there is no particle physics beyond the Standard Model, and as I have just said, there is no experimental evidence to the contrary, then all matter must have originated in this process somehow, in the early universe.

In the physics literature this mechanism of the generation of baryon and lepton is called *sphaleron baryogenesis. Sphaleron* comes from a Greek word meaning "ready to fall." This word is appropriate because the energy barriers that separate the different kinks in the vacuum can be pictured as an infinite series of wave peaks and valleys, with a valley corresponding to where a baryon or lepton would rest if it had no kinetic energy, and the peak, of height 10 TeV above this 0 of energy, being where the particle would have to be if it is to fall into another valley, which would cause another new baryon or lepton to form. The physicist who first wrote about sphalerons, and so got the privilege of naming them, imagined the particles on the top of the peak, where they were indeed "ready to fall."

In the early universe, the particles, especially the particles of the SU(2) field, were at so high an energy that the 10 TeV height of the energy peaks was easily overcome: practically all particles had an energy greater than a mere 10 TeV. Today, if one wishes to create or destroy matter by using this process, the energy to surmount the peaks must be obtained from elsewhere. The average kinetic energy of a particle at room temperature is only one-fortieth of a single electron volt. In Chapter 8, I shall argue that it should be possible to borrow sufficient energy from the other universes of the multiverse, as I described when I discussed quantum tunneling, to overcome the potential barrier of height 10 TeV. I shall term the physical system that accomplishes this temporary borrowing of energy from another universe of the multiverse to create or destroy baryons and leptons a *sphaleron field.*

It is interesting that, if this matter-creating process exists, it can be used to completely convert matter into energy. Without this process, one can get pure energy only if one interacts an equal amount of matter and antimatter: $e^- + e^+ \rightarrow 2\gamma$, where γ represents a photon, is allowed but complete annihilation of only matter is not. One cannot annihilate a hydrogen atom, for example: $e^- + p \rightarrow 2\gamma$ is not allowed. The barriers to pure mass-energy conversion are the two laws of lepton and baryon conservation. For the electron-proton reaction, we have +1 unit of lepton number and +1 unit of baryon number before the reaction, and if the two conservation laws hold, then these numbers must be the same after the reaction. There is no problem with the electron-positron reaction, since $(+1) + (-1) = 0$ both before and after.

The Standard Model process of creating matter violates both conservation laws but conserves $B - L$, where B is the baryon number and L is the lepton number. In this new process, it is possible to annihilate a hydrogen atom, since $B - L$ for the pair $e^- + p$ is $(+1) - (+1) = 0$. So if we could find a way to use this process, which must exist according to the Standard Model, we will have a powerful new energy source, which will convert 100 percent of the mass into energy instead of less than 1 percent, as in nuclear reactions.

It is this process that I shall suggest was the mechanism used in Jesus' Resurrection.

III

Life and the Ultimate Future of the Universe

LIFE ON EARTH IS DOOMED. THE ANNIHILATION OF ALL LIFE on this planet is an automatic consequence of the laws of physics outlined in the preceding chapter. The Sun is slowly but surely using up the nuclear fuel in its core. As the hydrogen in the Sun's core is converted into helium by thermonuclear fusion, the Sun gets steadily more luminous. In about 5 billion years, this increasing luminosity will cause the Sun to expand outward and engulf the Earth, thereby vaporizing the planet. Unless life leaves the planet before this occurs, life will die with the Earth.

This is a scientific prediction of the end of the world. But the end of the world had been predicted before modern science developed, in Jewish and Christian revelation. The visions of the destruction of the Earth in the Bible were predicted to occur in the "near" future—within decades or centuries instead of billions of years. And the end, it was said, would come rapidly. Speaking to his disciples, Jesus predicted a catastrophe of such magnitude that the human race would be obliterated unless he intervened directly, by "coming again in power." As we shall see in later chapters, the laws of physics suggest an end to human history in the "near" future—sometime in this century—in a manner strikingly similar to the end Jesus pictured to his disciples. But a close reading of this biblical passage and the Revelation of John indicate that this catastrophe will be limited to the planet Earth. The multiverse of universes will continue, as will life. As we shall see, the end of human history and the end of the Earth are required if life and the universe are to survive themselves.

Nevertheless, the opening prediction that, were life to remain on Earth, the Sun would destroy the entire biosphere, is still valid. What will actually happen is that in the future life will leave the Earth, expand through the cosmos, convert the entire universe into a biosphere, and convert the Earth (and Sun) into a "new heaven and a new earth." To understand how this will come about, and why the laws of physics confirm that it will, we have to understand what the laws of physics dictate for cosmology.

Cosmology and the Singularity

Cosmology is the study of the universe, and the multiverse, as a whole. Cosmology, then, is the study of reality on the largest possible scales. Modern physical cosmology began in 1917, when Albert Einstein (1879–1955) applied his newly discovered theory of gravity to the universe. Einstein realized that his equations implied the expansion (or contraction) of the universe in the large, but he was unable to accept this conclusion on philosophical and religious grounds. Einstein hated the idea of a personal God, who created the universe a finite time ago; he envisaged the universe as self-existent, an entity that had existed, and will exist, forever, unchanged, for infinite time. Such a universe was not allowed by his equations of gravity, so he modified them to a new set of equations that would allow an unchanging, eternal universe: the Einstein static universe.

Unfortunately for Einstein as a religious philosopher, other physicists quickly proved mathematically that his static universe was unstable: a tiny change—say moving a teacup from one side of the table to the other—would start the universe expanding. Within a few years, the American astronomer Edwin Hubble (1889–1953) discovered that the galaxies are, in the large, all moving away from one another. Einstein's original equations, unmodified by philosophical and religious considerations, were correct. Einstein later characterized modifying his original equations, adding a term called the *cosmological constant term*, as "the worst mistake of my life." I agree. Modifying a physical law simply because it has consequences unacceptable for philosophical or religious reasons is always a terrible mistake, the worst mistake a physicist, or any scientist, can possibly make. I shall not make that mistake in this book. As I emphasized at the beginning of the previous chapter, I shall accept the implications of the firmly tested laws of physics, whatever they happen to be. These firmly tested laws of physics I outlined in the previous chapter.

One of the implications of the laws of physics, an implication that most physicists find philosophically and religiously repugnant, is a necessary consequence of the expansion of the universe: it began a finite time ago—the latest measurements indicate 13.4 billion years ago[1]—in a *singularity,* where the laws of physics themselves do not apply. The laws of physics do not apply at a singularity because, as the initial singularity is approached from inside space and time, physical quantities such as the density of material go to infinity. The laws of physics, however, can govern only the behavior of finite quantities. In the words of the great cosmologist Sir Fred Hoyle (1915–2001), "The problem with a singularity is that not only do the known laws of physics not apply there, no possible laws of physics can apply there." Hoyle is completely correct; no possible laws of physics can control a singularity. Modern physicists *hate* the idea that something real could be beyond the power of the laws of physics. Almost as bad is the idea that the universe has existed for only a finite time.

In spite of most physicists' wishes, the universe began in a singularity 13.4 billion years ago. The reason the laws of physics require an initial singularity is not really hard to understand. The universe is now expanding, which means that, in the past, the galaxies were closer together than they are today. This means that the density of matter—recall that the density of matter is defined as the mass of matter divided by the volume in which this mass is contained—must have been greater in the past than it is now. The total mass making up a million galaxies is essentially unchanged over time, while the volume containing those galaxies was much smaller in the past. For example, say the galaxies are contained now in a volume of 1 million (in some units), while at a certain time in the past these million galaxies were contained in a volume of 100. The density now is 1 million divided by 1 million, or 1, whereas the density at the time in the past was 1 million divided by 100, or 10,000.

Now as one goes further and further into the past, the galaxies cease to exist (they were not formed until about 1 billion years after the universe began). But the matter making up the galaxies did exist, and the

These examples show that there are at least four possibilities for the actual topology of the universe: Euclidean space R^3, the topology $R^2 \times S^1$, the 3-torus, and the 3-sphere. The last two are finite in all directions and are called *compact* topologies. The first two are infinite in at least one direction and so are called *noncompact*. There are an infinite number of topologies in addition to these that the universe could have. What topology does the universe actually have?

Unfortunately, we can't answer this question by looking out into space. If the universe is curled up in one of its three dimensions, then the curvature is very small, because as far as we can tell, the universe seems to be flat Euclidean three-space. But this observation does not rule out the three other topologies, because it could just mean that we haven't looked at the universe on a sufficiently large scale. If we look at the surface of the Earth in all directions from the surface of a flat plain, or look at the surface of the ocean from a ship in the middle of the Pacific, the Earth looks perfectly flat. Our ancestors thousands of years ago thought the Earth was indeed flat. But we now know it is round, bending around on itself to form a 2-sphere.

Fortunately, using the laws of physics described in the previous chapter, we can determine the topology of the universe, because the laws of physics are completely consistent on only a very limited number of topologies. Only one, in fact. There are several ways to pick out this unique topology, but I have the space to mention only a few.

For example, in the presence of a weak gravitational field, a global distinction between matter and antimatter is possible only if the topology is compact. If matter cannot be distinguished from antimatter, this would in general lead to a catastrophic conversion of matter into antimatter, in contradiction with experiment. Second, only in universes that are spatially compact is it possible for event horizons to disappear. We shall discuss the problems with event horizons later in the chapter, where the term *event horizon* will be precisely defined. For now, let us merely note that event horizons are usually discussed in connection with black holes, for which event horizons form the surface. Hawking showed many years ago that if black hole event horizons were to form, then they would eventually violate unitarity, a central law of quantum mechanics, as I pointed out in the previous chapter.

Let us follow back into time this compact universe of ours. It is enormous in size now, but in the past, as the Initial Singularity is approached, its size goes to 0. Apply the Bekenstein Bound, which I described in the previous chapter, to the universe as the universe approaches 0 size. The mass-energy in a finite universe is some finite number, but the radius of the universe goes to 0. The Bekenstein Bound then says that the information content of the universe must go to 0 at the beginning of time. But if there were any variation in the distribution of matter in the beginning or any gravitational waves present in the universe, then this distribution of matter or gravitation wave would contain information. So the distribution of matter must have been perfectly regular sufficiently close to the Initial Singularity. Since gravitational waves are forms of curvature, their absence implies that, no matter in what direction we measure the curvatures, we get the same answer (in technical mathematical terms, we say that the sectional curvatures are constant).

Now the different compact topologies differ in various ways, but the only way we need to consider is how closed curves behave in them. Look at the two examples of compact two-dimensional spaces that I described earlier, the 2-torus (surface of a doughnut) and the 2-sphere (surface of the Earth). All such closed curves—circles—on the surface of the 2-sphere can be continuously shrunk to 0 size while remaining on the 2-sphere. This is not true of all circles on the surface of the 2-torus. If a circle happens to thread through the center of the doughnut, then it cannot be shrunk to zero size without passing through the body of the doughnut itself; that is, it cannot be shrunk to 0 size without leaving the surface of the doughnut. In technical mathematical language, the 2-sphere is said to be *simply connected,* while the 2-torus is not. There are actually two classes of circles on the 2-torus that cannot be shrunk to 0 size: the ones that pass through the hole, and the ones that circle the doughnut around the hole. This means that describing the topology of the doughnut requires more information to be given than does describing the sphere. The sphere is uniquely simple among the various compact topologies.

This fact means that the Bekenstein Bound will select out the 3-sphere as the only allowed topology, because any other topology will mean that there is irreducible information coded in the topology of the universe, whereas the Bekenstein Bound allows no information at all to exist in the universe at a sufficiently early time. Thus, the universe must be spatially a 3-sphere (there is a mathematical theorem that says the only simply connected compact space with constant sectional curvature is a sphere). We have incidentally solved, in the process of determining the universe's topology, an outstanding cosmology problem, the homogeneity problem (sometimes called the horizon problem), which poses the question of why the universe was so extremely regular everywhere in its early stages. The answer is, simply, that the laws of physics, specifically quantum mechanics in the form of the Bekenstein Bound, do not allow the universe to be irregular in its early history.

Another problem of cosmology is, Why is the universe so close to being flat? Why, if the universe is indeed a 3-sphere in its spatial topology, is it so large? Why is it so large that we cannot easily see that it is not flat? There are several ways to understand the huge size of the universe. The first is to realize that if the universe began at the Initial Singularity, the same must be true of the entire multiverse of universes. That is, at the Initial Singularity, the wave function of the multiverse—recall from Chapter 2 that this wave function codes the density of universes in the multiverse—must have been entirely concentrated on zero size, just as the single universe we find ourselves in was of zero size at the Initial Singularity. Recall from Chapter 2 the derivation of the uncertainty principle: as the size of the opening in the breakwater got smaller and smaller, the spread of the wave function coming out of the opening got broader and broader. In the limit that the size of the opening goes to zero size, the spread goes to infinity. Applied to the wave function of the multiverse, this means that, at all times after the Initial Singularity, the particular universe we find ourselves in is overwhelmingly likely to be of arbitrarily large size. So our universe, though finite in size and spherical in shape, is nevertheless very large and hence, like the surface of our planet Earth, very close to flat.

The universe on large scales is also very close to being "classical," which just means that, to a very high degree of approximation, we can completely ignore the effects of quantum mechanics on these scales in the present epoch of universal history. This is just another way to say that we can ignore the multiverse itself on these large scales and pretend that there is only one universe in existence, the particular universe we find ourselves in. The multiverse is still there, of course. It turns out that, mathematically, the requirement that the evolution of the universe is classical on the largest scales is equivalent to the requirement that the multiverse began at a single initial singularity: that is, the requirement that the wave function of the multiverse was initially concentrated entirely at a point. This implies, as I indicated before, that the universe we observe must be a very large sphere today. So there are two alternative ways of understanding why the universe is as large as it is observed to be.

The Ultimate Future of the Universe

As large as it is today, the universe is expanding, getting larger. What will be its ultimate future? Will the universe (and multiverse) expand forever, getting larger and larger without limit, or will it eventually stop expanding? To answer this question, we have to understand the matter content of the universe, because the matter determines the strength of the gravitational field in the universe on the largest scales, and it is the gravitational field strength that will determine the universe's future.

The Standard Model combined with the unique theory of quantum gravity, both described in the previous chapter, tell us what this matter is. The most important matter is the Higgs field, since, as pointed out in the previous chapter, its density near its minimum of potential is enormous, in magnitude some 10^{26} grams per cubic centimeter, compared with the measured density of all other forms of matter, only 10^{-29} grams per cubic centimeter. Notice what is implied by these numbers. We don't really seem to measure the Higgs field at all, because the density of everything in the universe, including the Higgs field, is actually fifty-four orders of magnitude less than the Higgs field mass-energy density itself. What can account for this huge difference?

One of the terms coming from the quantum gravitational theory accounts for this difference. It is called the *cosmological constant,* and the consistency of quantum gravity requires it to be present. It is interesting that this cosmological constant is the same term Einstein called "the worst mistake of my life." It was a mistake for Einstein to introduce this term not because it does not exist—it does—but because he had no experimental or theoretical reason for introducing it. In contrast, we have a very good theoretical reason for introducing the cosmological constant: it is necessarily present in the only consistent theory of quantum gravity. Quantum mechanics—specifically the linearity of quantum mechanics, which keeps singularities out of the laboratory and restricts them to the beginning and end of time, where singularities necessarily exist—requires the cosmological constant to be present.

The gravitational effect of the cosmological constant is, remarkably, almost exactly the same as the gravitational effect of the global Higgs field near its minimum of potential. As we shall see, the consistency of the Standard Model and the quantum theory of gravity also requires the true cosmological constant to cancel almost precisely the pseudocosmological constant arising from the Higgs field. In fact, they would exactly cancel if the Higgs field were at its true minimum. But recall from Chapter 2 that the Higgs field is not at its true minimum because more matter than antimatter was created in the very early universe. Only if this matter were annihilated by undoing the kinks in the vacuum would the real cosmological and pseudocosmological constants exactly cancel. The Standard Model tells us that the Higgs field today is very large in magnitude but negative in sign mass density, so the true cosmological constant must be very large and positive. Since the true cosmological constant has today a larger magnitude, we see a very small but positive cosmological constant after the effects of the true cosmological constant and the pseudocosmological constant are combined.

This small cosmological constant was first definitely detected in 1998, and it is called the *dark energy*. The adjective *dark* refers to the fact that we cannot see it via its generation of light (not surprising, since it generates no light); we can see it only by its gravitational effects. The noun *energy* is used to describe it rather than the noun *matter* because a cosmological constant necessarily acts gravitationally as if it had an enormous pressure associated with it. The effect of this pressure in a positive cosmological constant is to give rise to a repulsive gravitational force, causing the expansion of the universe to accelerate. In the present epoch of universal history, the dark energy makes up about 73 percent of the total mass-energy of the universe, ordinary matter makes up about 4 percent, and the remaining 23 percent is made up of what is called the *dark matter*. As is the case with the dark energy, the dark matter is called *dark* because it is seen only by its gravitational effects, not by its generation of light, as is the case with ordinary matter (the stuff generated in the early universe by the formation of kinks in the vacuum). The dark matter is called *matter* because, in contrast to the dark energy, it exerts no pressure. The Standard Model and the only mathematically consistent quantum theory of gravity leave possibilities for what this dark matter could be, and since the exact answer is not important for the subject matter of this book, I shall relegate the details of the dark matter to the physics journals.[3]

The essential fact we need to infer the future evolution of the universe is that the matter content of the universe consists of two basic sorts of "stuff": the dark energy, which acts gravitationally like a positive cosmological constant, and matter (of two sorts), which exerts no pressure in the present epoch. As the universe expands, the density of matter, and hence its gravitational effect, decreases inversely as the cube of the radius of the universe, since density is the ratio of the mass—for matter this is constant—to the volume enclosing the mass, and the volume is proportional to the cube of the size of the universe. Neither the mass density nor the pressure of the effective cosmological constant changes as the universe changes its size, and this is why it is called a *constant.* This means that, as the universe ages, the gravitational effect of the matter gets less and less, and the gravitational effect of the dark energy (effective cosmological constant) gets greater and greater. If the universe were to expand forever, its acceleration would get larger and larger. In the ultimate future of such a universe, Hubble's constant, which is a measure of the expansion of the universe, would approach a constant value.

In such a universe, life would die out. There are two reasons why an ever-accelerating universe is lethal. First, the acceleration would rip large-scale structures apart, and I showed in my previous book, *The Physics of Immortality,* that in the far future of an ever-expanding universe, life can survive only by expanding its reach over ever-larger regions. Second, *event horizons* develop in an ever-accelerating universe, and the essence of event horizons is that they prevent the communication of information from one part of the universe to another.

Event horizons are such important barriers to the continued existence of life that it is worth describing them in detail, and essential to understand how they arise in cosmology. A friend of mine, Wolfgang Rindler (now a professor of physics at the University of Texas at Dallas), coined the term *event horizon* in the 1950s in an important paper devoted to studying their properties. I think Rindler's term is exceedingly apt. An *event* is a specific location in space and time. Suppose you are reading this book at 9:00 P.M. on December 25, 2007, in your fifth-floor apartment in Washington, D.C. Four numbers describe this "event": three numbers giving your spatial location, and a fourth number giving the time. The word *horizon* refers to the fact that the speed of light restricts the events in the universe you can see. A person on the Moon, for example, would not be able to see you reading the book until a little more than one second after 9:00 P.M. An observer on a planet around the nearest star, Alpha Centauri, would not be able to see this event of you reading the book at 9:00 P.M. until four and a half years later. An observer on a planet in the Virgo cluster of galaxies would not be able to see this event until 60 million years later.

You might think that, just as all the observers mentioned so far would *eventually* see the event of you reading the book at 9:00 P.M. in Washington, this would be true for all observers, no matter where they were located in the universe. Rindler realized that this need not be true. Because of the expansion of the universe, and the fact that the farther a galaxy is from us the faster it is moving away, it takes much longer for a light ray to reach us than just the time needed for light to traverse the given galaxy's current distance from us. For example, if a galaxy were a billion light-years away from us now, then in order to find the time needed for light to reach us from that galaxy, we have to take into account the universe's amount of expansion over that billion years. The light will still be moving toward us from that galaxy after a billion years because of the universe's expansion.

It gets worse. Remember, not only is the universe expanding but the expansion rate is accelerating. When this acceleration is considered, it turns out that there are galaxies we will never see in the entire future history of an accelerating universe. Furthermore, there are galaxies we can see now that will eventually move and accelerate so far away from us that there will come a time when we will never again be able to see events in those galaxies in the remaining course of future history, even though this future history would be infinite. There is, in other words, a boundary between events in the universe that we will eventually be able to see and events that we will never, ever be able to see. This boundary is the event horizon. Just as the horizon on the surface of the Earth divides the part of the Earth we can see from the part we cannot see, the event horizon divides the universe into events we will eventually see if we wait long enough and events we will never see, even if we wait forever. In fact, Lord Martin Rees, the Astronomer Royal of England, has calculated that, if the currently observed acceleration were to continue at its present rate forever, all galaxies beyond the twenty-odd galaxies in the Local Group would eventually pass beyond our event horizon. The far future of such a universe would be a lonely place indeed.

Event horizons can arise by mechanisms other than the acceleration of the universe. If a star were to collapse to a very small size, to a radius of about 3 kilometers if the star had a mass equal to the Sun, then in a universe that existed forever, events inside this star could never be seen from outside the star in infinite time. The star would have become a *black hole.* By definition, the event horizon of the collapsed star is the surface (in spacetime) of the black hole. In this case, the event horizon is formed not by the acceleration of the universe but by the powerful gravitational field of the collapsed star, which is so strong that nothing, not even light, can escape.

Event horizons will also exist inside the black hole, an environment which closely resembles the end history of a universe that collapses to a final singularity. What happens is that the matter in the interior of a black hole, and possibly the matter in a collapsing universe, can collapse inward so fast that light will not have time to cross the distance from one side of the collapsing region to the other before a singularity is reached, stopping light and everything else.

An event horizon, however it arises, is always a surface defined by the paths of light rays. In black holes, the light rays whose paths define the event horizon are trying to escape the collapsing star but are instead so strongly attracted by its gravitational field that they remain forever at a constant distance from the center of the star. Light rays just inside the event horizon are more completely caught. Even the outgoing light rays are pulled irresistibly toward the center of the collapsing star. Imagine the surface of a 2-sphere just inside the event horizon. Imagine covering both the inside of the sphere and the outside with flash powder and then setting the powder off. There would be two sets of light rays from the two flashes, one moving inward and one outward. In the instant after the flash, the wave front of the light inside the sphere will form a smaller sphere moving toward the center of the sphere, and the wave front of the light outside the sphere will form a larger sphere moving away from the sphere.

At least this is what would happen if the sphere were outside a black hole. If the sphere in question were just a little less than 3 kilometers in radius and inside a black hole with a mass equal to our Sun's, then both the inward-going light front *and* the outward-going light front would be smaller! The outward-moving light front is smaller because of the powerful attraction of the black hole's gravitational field. A 2-sphere for which both the inward-moving and outward-moving light fronts are smaller is called a *trapped surface.* If the universe were to expand forever, all trapped surfaces would be hidden behind the event horizons of black holes. This being the case, astrophysicists use the existence of trapped surfaces as the hallmark of a black hole. They cannot actually see a trapped surface, of course, but they infer its existence by observing the amount of mass in a volume in space. If the volume is sufficiently small relative to a given amount of mass, the theory of gravity guarantees that a trapped surface exists. An event horizon cannot be measured, even indirectly, because it is defined not by what is happening now but by what will happen over the entire course of future history. Obviously, we cannot do an experiment now over the infinite future. But *if* the universe *were* to exist for infinite time, there would be an event horizon, and the (almost) observed trapped surface would be just inside it. Remember the italicized qualifications; they will be seen later to be very, very important.

In 1974, Stephen Hawking made a remarkable discovery: black holes evaporate. That is, he discovered that black holes cannot exist forever because they would radiate away their mass at a rate that would be inversely proportional to the square of their mass. So as a black hole's mass is radiated away, the mass would get smaller, which in turn would mean that the black hole would radiate its mass away even faster, and so on. In the end, after a black hole had radiated away almost all its mass, it would appear to radiate away the rest in one huge explosion.

There is a problem with this picture, as Hawking was quick to point out. A light ray lying in an event horizon cannot have a future end point. If it did, this end point, by definition, could be seen from outside the black hole. But if this end point can be seen, then any point of the light ray could be seen, which means that the light ray is not lying in the event horizon, contrary to assumption. But if the black hole came to an end by radiating away all its mass, what would happen to its event horizon? The light rays that generate the event horizon cannot end in a point in spacetime, but they must nevertheless come to an end, since the black hole has come to an end. Therefore, the event horizon must end in a singularity that lies *inside* spacetime rather than on its boundary. Such a singularity, Hawking pointed out, would violate unitarity. This is very bad, because unitarity, as I discussed in Chapter 2, is a fundamental principle of quantum mechanics. Over the past thirty years, numerous proposals have been made to solve this "black hole information problem," but the laws of physics permit only one, suggested by Hawking himself in a famous lecture given in Dublin in the summer of 2004.

Event horizons cannot exist. If they exist in any form, then the particular lethal form of black hole event horizons will also exist. But it is a theorem of relativity that event horizons can totally vanish only if the universe is spatially compact, which is to say, spatially closed. Given spatial compactness, I have already remarked that other physical laws will force the universe's spatial topology—its shape—to be a three-dimensional sphere.

The nonexistence of event horizons is so important that it is worth exploring in some detail how event horizons can disappear in universes that are compact, or spatially finite. As I pointed out earlier, if the universe expands with sufficient rapidity, as it would if it were to accelerate forever, then event horizons form because, in effect, the universe expands so fast that light cannot keep up. Event horizons can also form if the universe collapses so fast that, once again, light cannot keep up. The final singularity is reached before light has a chance to travel the distance between one spatial location and another.

Suppose that the universe picked a happy medium and expanded or collapsed more slowly than in either of these extreme event horizon cases. There would still be black hole event horizons in the expanding case, no matter how slowly the universe expanded. However, in the collapsing universe, and only in the collapsing universe, it is possible for event horizons to be completely absent. Let us first understand how the event horizons of cosmology can vanish and then tackle the black hole event horizons in this closed universe.

An event-horizon-free closed universe is pictured in Figure 3.1. The convention that we measure distance in light-years and time in years is used in this figure, so that the paths of light rays are the lines oriented at 45 degrees with respect to the vertical time line.

In this universe, the approach to the final singularity is sufficiently slow that any light ray can circumnavigate the entire universe an infinite number of times before the final singularity is reached.[4] Thus, no matter how close to the final singularity an event is located, a light ray from that event can reach every other spatial location in the entire universe not just once but an infinite number of times. Notice that in Figure 3.1 the singularity is represented by a single point.

The final singularity in a universe without event horizons is represented by a single point because it *is* a single point. To see this, we first have to understand how a singularity can have a structure more complicated than a single point. The English mathematician Roger Penrose proposed that we put a topology on a singularity in the following way.

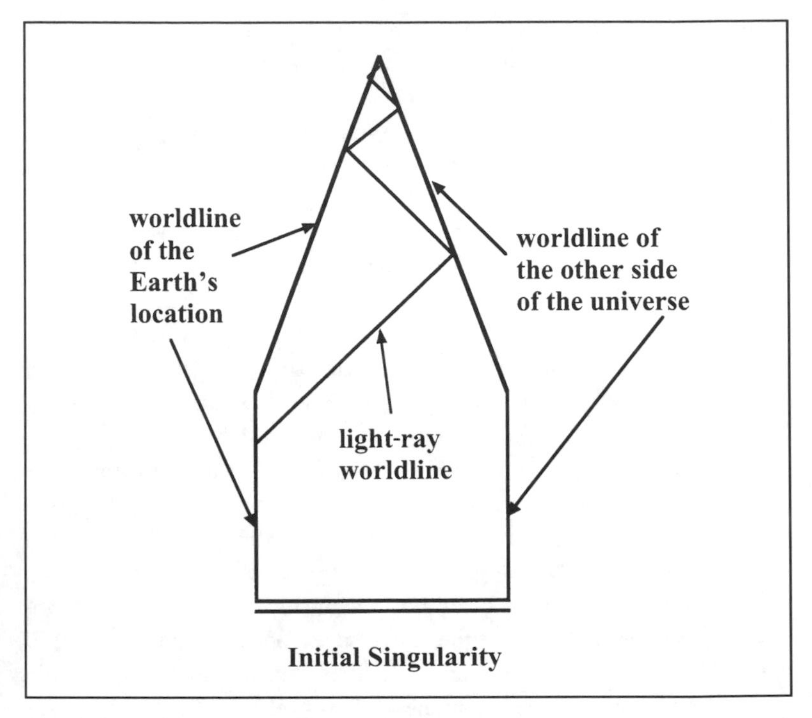

Figure 3.1. Penrose diagram of a closed universe without an event horizon. A light-ray worldline is shown, and it circumnavigates the universe an infinite number of times. All events are visible to all observers in this universe.

Consider each curve that represents a history of a particle traveling at the speed of light or less: such a curve is called a *causal* curve because it can carry information into the future. Consider a causal curve that has no end point in the future in spacetime. Such a curve has no future end point either because it goes on forever in a universe that itself goes on forever or because it ends at the future singularity when the rest of the universe ends.

Consider now a singularity in a 3-sphere universe that collapses so fast it develops event horizons.[5] In this case, there will be causal curves which have different sets of past events that can influence them. By looking at all causal curves, we can form a collection of all distinct sets of past events that can influence these curves. Some future endless causal curves that hit a final singularity will define the same set of past events that can influence them. These curves will be said to define the same "point" on the singularity. Other causal curves will define different sets of past events, and these will be said to define different "points" on the final singularity. Penrose proposed that the set of all distinct "points" on the final singularity actually forms the final singularity. More generally, the set of all of these future "points" defines a spacetime's *c-boundary.* The letter *c* stands for "causal."

In Figure 3.2, a Penrose diagram of a 3-sphere closed universe in which radiation is the dominant component of the effective energy content is pictured. As always, the paths of light rays are oriented at 45 degrees with respect to the vertical time direction.

In such a universe, each causal curve corresponding to a spatiotemporal location with unchanging spatial position defines a distinct point on the future c-boundary, and there are no other points on the c-boundary. This means that the topology of the final singularity in this universe

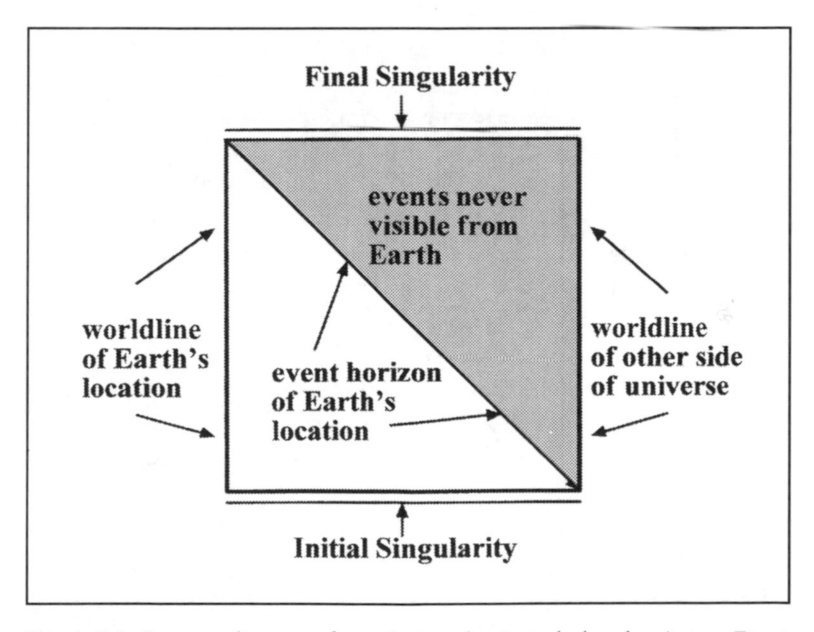

Figure 3.2. Penrose diagram of a radiation-dominated closed universe. Event horizons necessarily exist, preventing different regions of the universe from ever communicating with one another. The darkened region—half of the spacetime—are the events forever invisible from the Earth's location. Any light ray sent out toward the Earth from any point in the darkened region would hit the final singularity before reaching the Earth. The singularities are denoted by double horizontal lines.

is not a single point but instead is the same as the topology of a closed universe itself; the c-boundary is a 3-sphere. A closed 3-sphere universe that accelerates forever has exactly the same c-boundary topology—a 3-sphere—although in this case the future endless causal curves are infinite in length.

A universe with no event horizons, by contrast, has only a single point for its future c-boundary, because in the absence of event horizons, all causal curves define the same past set: necessarily all events in spacetime can be seen by all observers sufficiently close to the final singularity. This single-point singularity of a universe without event horizons I have called an *Omega Point.* I have shown earlier that the mutual consistency of relativity theory and quantum theory requires that the actual universe be an Omega Point spacetime.

Life and the Omega Point

The theory of such a universe, the Omega Point theory, I developed at length in my earlier book, *The Physics of Immortality,* though in that book I derived the existence of the Omega Point singularity from the requirement that life exist forever. I have since improved my argument: the existence of the Omega Point singularity is an automatic consequence of the most fundamental laws of physics, specifically quantum mechanics and relativity. Life has not come into the argument.

But life indeed has an important role to play in the cosmos. If to quantum mechanics and relativity we add further physical laws, specifically the Standard Model of particle physics and the Second Law of Thermodynamics, we can use physics to deduce this role. I shall show that the mutual consistency of all these laws *requires* two additional things. First, life must survive to the very end of time. Second, the knowledge possessed by life must increase to infinity as the end of time is approached. I do not *assume* life survives to the end of time. Life's survival follows from the laws of physics. If the laws of physics be for us, who can be against us?

But before I prove that the laws of physics *require* life to survive, let me first show that it is *possible* for life to survive. To survive for infinite experiential time, life requires an unlimited supply of energy. That is, the supply of available energy must diverge to infinity as the end of time is approached. Nevertheless, conservation of energy requires the total energy of the universe to be constant. In fact, Roger Penrose has shown that the total energy of any closed universe is *zero!* The total energy is zero now, was zero in the past, and will be zero at all times in the future. One might wonder how this is possible. After all, we are now receiving energy from the Sun, we are using food energy as we read this, and we can extract energy from coal, oil, and uranium. Energy, in other words, seems to be nonzero.

However, the forms of energy just listed are not all the forms of energy in the universe. There is also gravitational energy, which is negative. This gravitational energy can have spectacular effects, as we shall see when we discuss hypernovae in Chapter 6. If we were to add all the positive forms of energy—radiant energy; the stored energy in coal, oil, and uranium; and most important, the mass-energy of matter—to the negative gravitational energy, the sum would be 0. This means that if we can make the gravitational energy even more negative, the positive energy, that is, the energy available for life, necessarily increases, even though the total energy in the universe stays 0. The key property of energy that must always be kept in mind is that it transforms from one form to another. The Nobel Prize–winning physicist Richard Feynman has emphasized this.[6] Once we realize that gravitational energy can be transformed into available energy, we understand where life can find the unlimited available energy it needs for survival: life must make the total gravitational energy approach minus infinity.

Life can do this only if the universe is closed and collapses to zero size as the end of time is approached. If the universe is closed and collapses to zero size, the total gravitational energy goes to minus infinity because the gravitational energy of a system is inversely proportional to the size of the system. I have shown in my book *The Physics of Immortality* that life can extract unlimited available energy from the collapse of the universe.

Now let me outline the proof of the two claims I made about life. I have given an outline of why the laws of physics require the universe to be spatially closed and without event horizons.[7] I should like to add another reason why the Final Singularity must be an Omega Point. If the Final Singularity were to be accompanied by event horizons, then the Bekenstein Bound would force all the microstate information in the universe to go to 0 as the universe approaches the Final Singularity. But the microstate information going to 0 would imply that the entropy of the universe would have to go to 0, and this would contradict the Second

Law of Thermodynamics, which says that the entropy of the universe can never decrease. But if event horizons do not exist, then the Bekenstein Bound allows the information in the microstates to diverge to infinity as the Final Singularity is approached. Conversely, *only* if event horizons do not exist can quantum mechanics (the Bekenstein Bound) be consistent with the Second Law of Thermodynamics. Therefore, event horizons cannot exist, and by Seifert's theorem, the nonexistence of event horizons requires the universe to be spatially closed.[8] In Penrose's c-boundary construction, a singularity without event horizons is a single point.[9] This is what I have called the Omega Point.

It is important to realize that "astrophysical" black holes can exist in a spacetime that ends in an Omega Point. This is counterintuitive, because black holes are defined by the presence of event horizons, while Omega Point spacetimes, by definition, have no event horizons. The key fact to keep in mind, however, is that no astronomer has ever seen an event horizon. What astronomers have seen are regions whose gravitational fields are so strong that it is reasonable to infer the existence of trapped surfaces in them. It is a theorem of relativity that if the universe were to expand forever, then these trapped surfaces would be forever shielded from our view by event horizons. *But it is an inference, an invalid inference, that trapped surfaces imply event horizons.* We could validly infer the existence of event horizons from trapped surfaces only if we knew experimentally that the universe expanded forever, and we could know this experimentally only if we had existed to the very end of the universe. Obviously, we have not observed the universe for infinite time. Therefore, we cannot say that our observation of trapped surfaces means event horizons exist.

On the one hand, if the universe were to evolve into the Final Singularity without life being present to guide its evolution, then the nonexistence of event horizons would mean that the universe would be evolving into an infinitely improbable state. Such an evolution would contradict the Second Law of Thermodynamics, which requires the universe to evolve from less probable to more probable states. But if life is presently guiding the evolution of the universe into the Final Singularity, then the absence of event horizons is actually the *most* probable state, because the absence of event horizons is exactly what life requires in order to survive, as I calculated in detail in the "Appendix for Scientists" of *The Physics of Immortality.* In other words, the validity of the Second Law of Thermodynamics *requires* life to be present all the way into the Final Singularity, and further, the Second Law requires life to guide the universe in such a way as to eliminate the event horizons.

Life is the only process consistent with known physical law capable of eliminating event horizons without the universe evolving into an infinitely improbable state. Exactly how life eliminates the event horizons is described in my earlier book. Roughly speaking, life nudges the universe so as to allow light to circumnavigate the universe first in one direction, and then in another. This is done repeatedly, an infinite number of times. There are thus an *infinite* number of circumnavigations of light before the Omega Point is reached. If we were to regard a single circumnavigation as one tick of the "light clock," there would be an infinite amount of such time between now and the Omega Point. An even more physical time would be the number of experiences life has between now and the Omega Point. This "experiential time"—the time experienced by life in the far future—is the most appropriate physical time to use near the Omega Point. It is far more appropriate than the human-based "proper time" we now use in our clocks. As we shall see, experiential time is infinite.

The strongest evidence for the acceleration of the universe comes from measurements of perturbations of the cosmic microwave background radiation. These measurements show that the universe is within 1 percent of being flat, as I predicted in *The Physics of Immortality,* and that these perturbations follow the scale-invariant Harrison-Zel'dovich spectrum, as I also predicted in that book. Many believe that these are unique predictions of inflation. This is false. Alan Guth's first paper on inflation appeared in 1981, but Edward Harrison's paper on the scale-invariant density fluctuation spectrum appeared in 1970, and Yacob Zel'dovich's paper in 1972.[10] If their spectrum were a unique consequence of inflation, then it would have been impossible for Harrison and Zel'dovich to have proposed the scale-invariant spectrum before inflation was invented. Instead, they proposed the scale-invariant spectrum a decade before Guth. The scale-invariant spectrum is the only perturbation spectrum globally consistent with a spatially flat metric, which, as Einstein and Willem de Sitter argued in the 1930s, is the cosmology we should prefer in the absence of proof that the universe is closed or open. Like Roger Penrose and my mentor John Wheeler, I reject the idea of inflation, because inflation assumes the existence of a force field (the inflation field) that has never been seen in the laboratory. Further, inflation has made no predictions that cannot be obtained from forces that have been observed in the laboratory. As I stated earlier, I follow Galileo and assume that the same forces act in the terrestrial and extraterrestrial realms.

If the universe is spatially flat, if the Hubble constant is 70 kilometers per second-megaparsec (as the most recent measurements suggest), and if the universe is 13.4 billion years old, as the latest observations of the cosmic microwave background radiation indicate, then necessarily the universe either is now accelerating or has accelerated in the past. The distribution of galaxies can best be explained by assuming the universe is accelerating now, and observation of the high-red-shift galaxies strongly suggests that the acceleration is driven by a positive cosmological constant.

When I wrote *The Physics of Immortality,* in 1994, it never occurred to me that acceleration would happen in the expansion phase of universal history. I did expect acceleration to happen in the collapsing phase of the universe, and it is this acceleration that allowed me to predict the value of the Higgs boson and top quark masses to be 220 ± 20 GeV and 185 ± 20 GeV, respectively. (The current value for the top quark mass, first measured the month after my book appeared in Germany, is 174 GeV. The Higgs has not yet been detected, but the current lower bound to its mass is 114 GeV.) The reason I never considered the possibility that the universe could accelerate in its expanding phase is that, if the acceleration were to continue forever, life would be wiped out, and the Omega Point would never come into existence. As I showed earlier, this would contradict unitarity by black hole evaporation. If the Standard Model of particle physics is true, then there can be only one cause of acceleration of the universe, namely a positive cosmological constant. As I pointed out in *The Physics of Immortality,* the Standard Model says the universe is in a vacuum of the Higgs field, and it says this vacuum would act today as a very large *negative* cosmological constant. If this vacuum were not currently canceled by a positive cosmological constant, the universe would collapse into a final singularity in a fraction of a second. Hence a positive cosmological constant must exist. I thus assumed in my earlier book that the Higgs field is in its absolute vacuum state today, where we would expect the positive cosmological constant to cancel the Higgs vacuum energy precisely.

But suppose the Higgs field is not in its absolute vacuum state. In such a case, the Higgs vacuum energy would only partially cancel the positive cosmological constant. The uncanceled part of the positive cosmological constant would cause the universe to accelerate when the matter density dropped low enough. But if the Higgs field is not in its absolute vacuum, there must be a mechanism to cause this, and also to allow the Higgs vacuum to relax to its absolute vacuum so that unitarity will not be violated.

The Standard Model provides such a mechanism, which I discussed in the last section of the "Appendix for Scientists" in *The Physics of Immortality*. I have described it in a more popular way in Chapter 2 of this book. This mechanism is the creation-destruction of baryon number by electroweak quantum tunneling. (Baryons are the heavy particles made up of quarks. Examples are neutrons and protons.) In my earlier book, I pointed out that this mechanism would be ideal for propelling interstellar spacecraft, but I did not discuss its implications for the Higgs vacuum, a serious oversight on my part (and an oversight that invalidates the second part of my fifth prediction on page 149 of *The Physics of Immortality*). If the Standard Model is true—and *all* experiments conducted to date (e.g., Wilczek and Quinn)[11] indicate that it is—then the net baryon number observed in the universe must have been created in the early universe by this mechanism of electroweak quantum tunneling. If the baryons were so created, this process necessarily forces the Higgs field to be in a vacuum state that is not its absolute vacuum. But if the baryons in the universe were to be annihilated by this process, say by the action of intelligent life, then this annihilation would force the Higgs field toward its absolute vacuum, canceling the positive cosmological constant, stopping the acceleration, and allowing the universe to collapse into the Omega Point. Conversely, if this process does not annihilate enough baryons, the positive cosmological constant will never be canceled, the universe will expand forever, unitarity will be violated, and the Omega Point will never come into existence. Only if life makes use of this process to annihilate baryons will the Omega Point come into existence.

It is not enough to annihilate some baryons. If the laws of physics are to be consistent over all time, a substantial percentage of all the baryons in the universe must be annihilated, and over a rather short time span. Only if this is done will the acceleration of the universe be halted. This means, in particular, that intelligent life from the terrestrial biosphere must move out into interstellar and intergalactic space, annihilating baryons as they go. So it must be "easy" to master the baryon-annihilation process; it should, in other words, be possible to carry out the process on a small scale. The annihilation process will in this case provide the means to traverse interstellar and intergalactic space.

Interstellar travel requires a rocket whose exhaust particles travel at very close to the speed of light, as I pointed out in the "Appendix for Scientists" in *The Physics of Immortality*. In this earlier work, I considered the only process known at the time for producing a light-speed exhaust, matter-antimatter annihilation, but with the antimatter carried as a fuel aboard the rocket. The baryon-annihilation process provides a much better system; in effect, it provides the ultimate rocket propulsion mechanism. All rocket scientists have long known what the perfect rocket would be like. First, the exhaust particles would travel out the rear of the rocket as fast as possible. In view of the constraint of relativity, this is the speed of light. Second, the exhaust particles should interact with other forms of matter as little as possible. Think of a typical rocket launch today. The exhaust gases are so hot and vast that humans observing the launch must be kept miles away. And this is using chemicals such as kerosene and liquid oxygen as the fuels. Rocket scientists have long wished to use nuclear energy as the fuel, but they know this is impossible because the exhaust would be radioactive. That is, a nuclear exhaust would interact massively with other matter, particularly biological matter.

Instead of using chemicals or nuclear reactions to propel a rocket, let us consider what the baryon-annihilation process can provide. Recall from Chapter 2 that the process will allow us to annihilate baryons and leptons, provided that the difference between the baryon and lepton numbers is unchanged. Thus, a hydrogen atom can be annihilated because both the proton and the electron in the hydrogen atom have +1 unit of baryon number and lepton number respectively, and $(+1) - (+1) = 0$. But recall that antimatter has a negative baryon or lepton number. A neutrino has +1 unit of lepton number (and 0 baryon number), and its antiparticle, the antineutrino, has -1 unit of lepton number. Thus the lepton number of a neutrino-antineutrino pair has a lepton number of $(+1) + (-1) = 0$. So the baryon-annihilation process would allow an annihilation of a hydrogen atom into a neutrino-antineutrino pair. It would also allow the annihilation of a neutron into a neutrino and two antineutrinos. Since all atoms are made up of proton-electron pairs and neutrons, it follows that the baryon-annihilation process allows us to convert all atoms into combinations of neutrinos and antineutrinos. This means that, in principle, a rocket whose exhaust consists entirely of neutrinos and antineutrinos, and whose energy source for these neutrinos and antineutrinos is any sort of ordinary matter, can be made.

This is the ultimate rocket. Neutrinos and antineutrinos have very close to zero mass, so they travel almost at the speed of light. Furthermore, they have very little interaction with matter: a lead shield would have to be a light-year in thickness to stop half of the neutrinos passing through it. Thus, to human eyes the exhaust of a neutrino-antineutrino rocket would be completely invisible, and it would be harmless to the environment. A neutrino-antineutrino rocket engine could be used to power a family aircraft, a future replacement for the family car. Not only would the family vehicle be able to fly, but it would not need expensive gasoline. Its energy source would be the conversion of any ordinary material into the energy in the neutrino-antineutrinos. Such a vehicle is illustrated in the final scenes of the science fiction movie *Back to the Future*: Doc, the inventor of a time machine built into an automobile, returns from the future with the car now powered by garbage dumped into a hopper on the car, and propelled by a rocket engine whose exhaust does not damage the immediate surroundings. The laws of physics do not allow us to build a time machine, but they will allow—indeed they will require—us to build a small car-aircraft powered by garbage.

The baryon-annihilation process can be used to provide energy for any purpose we wish: by converting atoms into photons rather than neutrinos, baryon annihilation can be used to provide electrical energy. This process of complete matter conversion into energy is the ultimate source of energy in the expansion phase of universal history (gravitational energy being the ultimate source of energy in the collapsing phase, as I have already described). Once we have mastered the techniques of baryon annihilation, oil, coal, hydroelectric, and nuclear forms of energy will be obsolete.

Unfortunately, this revolutionary new form of energy will bring with it a great danger: the possibility of revolutionary new weapons. Baryon number conservation restricts the explosive power of nuclear weapons, enormous as this power is. Roughly, only 1 percent of the mass-energy contained in uranium (fission bomb) or lithium deuteride (fusion bomb) is available for an explosion if the baryon number must be conserved. But the baryon-annihilation process allows 100 percent of the mass-energy of ordinary materials to be converted into the energy of an explosion. Worse, since any material can be converted completely into energy, there is no rare material constraint to building a bomb nor any limit on the bomb's size. Today, a potential terrorist must first gather about 56 kilograms of enriched uranium or a substantially smaller amount of plutonium before he can destroy a city.[12] With the baryon-annihilation process, a few kilograms of garbage, converted entirely into energy, will annihilate a city. A few million tons of nickel and iron in the Earth's core, converted completely into energy, would blow the entire planet apart (a few million tons of energy—remember $E = mc^2$—is the gravitational binding energy of the Earth). If such a disrupter bomb were constructed and detonated, we would have a new asteroid belt in our solar system. Thus, the baryon-annihilation process carries not only great benefits but great dangers. But never forget that we, or our descendants, *must* eventually learn to use this process in order to stop the acceleration of the universe. If we never learn to use this process, then all life will die out, and the laws of physics will become inconsistent. The laws of physics are absolutely firm; therefore, our descendants will eventually learn to use this process.

A revolutionary new energy source is not the only challenge we humans will face in this century or the next. We shall also face the development of artificial intelligence (AI) and human downloads. An *artificial intelligence* is a computer program capable of duplicating all human intellectual activity, but on a computer rather than in a human nervous system. Notice that it is better to consider an AI as a program rather than a computer, just as a human being is not a brain but a *soul,* which is an immaterial entity generated by the activity of neurons in a human brain. A computer program is equally immaterial: it is a pattern, an active pattern, in matter rather than a form of matter itself. St. Thomas Aquinas, following Aristotle, defined the human soul as the "form of activity of the body." In the Aristotelian physics presupposed by Aquinas, a "form" was a pattern imposed on matter rather than matter itself. Aristotle used the particular shape of a statue as his example of a "form." A computer program, which is a series of numbers on a CD or hard disk, is exactly the same thing. So following Aristotle and Aquinas, we can regard the human soul as a form of computer program, a program running on a wet computer we call the human brain.

Many Christians have an unfortunate tendency to imagine the soul as a sort of white, ghostly substance that permeates the human body, to be released at death. This is a terrible mistake. If the soul were a substance, it would indeed *be* a substance, and hence material, contrary to the correct Christian claim that the soul is immaterial. Worse, this ghostly material view of the soul is heretical, for it suggests that there is a separate "spiritual" but substantial world parallel to this material world that we see. As I shall discuss in Chapter 5, this is the Gnostic worldview, not the Christian worldview. Worst of all, thinking of the soul as a ghostly substance prevents us from understanding Who God really is, how God can be Three Persons in one God, and how the Incarnation worked as a matter of physics. If the soul were a ghostly substance, it would presumably be subject to laws outside the laws of this universe, and this "spiritual universe" would be the universe of real interest. The physics of this universe would be of no fundamental interest, contrary to the message of Genesis, Chapter 1, which repeatedly asserts the goodness and importance of the material universe.

In *The Physics of Immortality,* I calculated how much power—speed and memory—a computer would have to have in order to run a program equal to the "soul program" currently being run on each of the 6 billion human brains on the Earth. Moore's law governs the advance of computer hardware: computer speed doubles every eighteen months. Moore's law indeed has held good over the period since my previous book: supercomputers now have a speed several times the processing power of the human brain, and we can expect desktop and laptop computers to equal today's supercomputer speeds within a decade or two. So everyone should expect to have on his or her desk or lap a computer with processing power equal to that of the computer in his or her head by the year 2025.

Why don't we have computers now that can think at the human level? Because, although we have the hardware to equal the human brain, we lack the software: we do not have a program to equal the human soul. We don't know how the intelligence program in the human brain works. Our current AI programs are laughably simple, gigantically less complex than the human soul program. Think of the schedule programs you "talk to" over the telephone when you make an airline reservation or find out when Uncle John's plane is landing. These AI programs can understand simple airline-related questions. But it is ridiculous to say you talked to the program. These programs are much too simple to have a true conversation with, as you can with another human being. At the moment, computer programmers are missing some crucial feature in the human-soul program, the key feature that allows a human baby to generate intelligence by an age between two and five. Computer programmers are missing the key that will allow them to duplicate the human mind in a silicon machine just as physicists are missing the key that will allow them to duplicate the baryon creation-annihilation process that operated in the early universe to create all the matter we now see. We can expect these two problems to be solved in the same century, possibly in the same decade or the same year.

It is even possible that solving the baryon-annihilation engineering problem would automatically lead to an AI program. The human brain's AI program is a very efficient program; it is likely programmers could create a hugely inefficient AI program if they had computer hardware gigantically more powerful than the human brain. And they would have such a gigantically powerful computer if a quantum computer could be made to work. A quantum computer uses coherence between the universes of the multiverse to share effectively the information in a program between universes. This is a multiverse version of the parallel processing that all supercomputers use in a single universe: rather than running the entire program in a single place in the processing center, the program is split up into subprograms, which are run in separate processing chips in different locations. As each chip completes its share of the program, the results are pulled together into a single location.

A quantum computer does the same thing, only across the multiverse. This process can work because the identical analogues of ourselves are necessarily interested in running identical programs at the same time. In effect, different parts of the program are run in different universes, and the final result is shared by all the analogues. The power of the quantum computer arises from the fact that the number of "distinct" universes increases exponentially with the number of atoms in each universe. The word *distinct* is in quotation marks because the universes differ only in the particular subprograms being run in them. On the human level, they are identical. Recall from Chapter 2, when we were discussing the Bekenstein Bound, I pointed out that the information content of a system is defined as the logarithm of the number of distinct quantum states. The number of states is thus the exponential of the number of quantum states, and it is this exponentiation that is being exploited in the quantum computer. The number of distinct universes is the number of states.

The engineering problem preventing the manufacture of a practical quantum computer is maintaining quantum coherence over more than a few atoms. Quantum computers using up to ten atoms to store information have been made, but these are quite useless, because 2 raised to the tenth power is only 1,024. That is, this quantum computer can store only 1,024 bits of information. A laptop computer can typically store 40 gigabytes, or about 400 billion bits, of information, so indeed a ten-atom quantum computer cannot compete with the average laptop. A quantum computer with 100 atoms can store 2^{100} bits, or about 10^{29} bytes, which is 10^{20} gigabytes of information. This is 100 million trillion gigabytes. A quantum computer using 100 atoms would give the average laptop serious competition. The general feeling among quantum computer researchers, a feeling I share, is that, if we could figure out how to maintain quantum coherence—allowing the universes of the multiverse to be "aware" of one another—at the 100-atom level, quantum coherence could be scaled up to trillions of atoms.

Atoms have energy levels that differ by a few electron volts. Quantum coherence among a trillion atoms would allow the atoms to concentrate the energy differences of the levels on a single atom, and this would be 10 TeV, the amount of energy needed for the baryon-annihilation process to go forward. So indeed the engineering insight needed to make a practical quantum computer is exactly the same as that required to annihilate baryons, to create the ultimate rocket and the ultimate energy source in the universes-expanding phase. I therefore expect the two problems to be solved about the same time, by the same technique.

A quantum computer would also be a nanocomputer, and thus would make nanotechnology practical. *Nanotechnology* is based on the atomic scale: machines that are made up of tens or hundreds of atoms instead of the many trillions of atoms that make up the smallest machines today, the basic memory units in a computer's memory chip. The word *nanotechnology* comes from *nanometer,* that is, one-billionth of a meter, which is the size of a typical atom. Once the quantum coherence problem is solved, it should be possible to construct nanocomputers to control the nanomachines. Such machines could be used for construction at the atomic level: assembling devices atom by atom. Some have expressed the fear that such machines could escape human control. Michael Crichton's 2002 novel *Prey* is based on such a nanomonster. But even in Crichton's novel, the nanomachine became an out-of-control monster only when it was given the power of self-reproduction, which allowed it to grow exponentially, like a disease germ. With self-reproduction capability, any machine can grow into a threat, or at least grow beyond human control. But since such self-reproducing machines will become technically feasible when quantum computers are technically feasible, we can expect to face this nanotechnology problem sometime in this century, about the same time as we are faced with quantum computers and with baryon-annihilation energy sources and weapons.

Human downloads are another current dream of computer theorists that would be made possible by the development of quantum computers. I mentioned that although we have hardware sufficient to run a computer program capable of equaling a human mind, we lack the software. We do not have a clue as to how the human brain generates a human mind. However, if one has sufficient computer power, such knowledge is not necessary to make an intelligent computer program. Every human can think at the human level, obviously. So map the entire contents of the human brain, including the location and state and connections of every neuron, into the memory of a computer. Not only this, but map the entire human body, and a suitable human environment, into a computer, and let the computer simulate a human being interacting with his or her environment. Such would be a human download: a complete human being converted into a computer simulation. Creating a human download is not technically feasible, mainly because we have no means of measuring the state of every neuron in the brain. But with nanotechnology guided by quantum computers, tiny robots could be injected into a human brain to map out the location and state of every neuronal connection. With enormous computer power, the simulation would be indistinguishable from a real human in a real environment, at least from inside the simulation.

These programs will be different in an essential way from the computers that most people are familiar with. In order to be people, these programs will of necessity have free will, and only a very few computers have been built to have free will. The desktop and laptop computers of daily life obviously do not have free will. Or perhaps I should say they have been designed not to have free will, even though they occasionally manifest free will when they do what *they* want to do rather than what we want them to do. This is a design flaw; we want our machines to do what *we* want them to do rather than what *they* want to do. However, creativity requires free will, and the machines that can simulate a human being will have to allow free will in the human simulations. I shall discuss how such a machine can be built in Chapter 11.

Once transformed into a human download, a simulated human could interact with the normal human world on the real Earth. For example, a simulated human could speak in the simulated environment. This simulated sound could be picked up by a simulated microphone, which would be programmed to generate exactly the same speech in a real microphone in the real world. The telephone everyone uses every day does exactly this. If you are speaking to someone you have never met, how do *you* know it's not a computer, or simulated human, you are talking to? The telephone voice generated by AI programs for the airline schedules is fairly human sounding, provided that you don't ask questions beyond its very limited capabilities. With a human download, you could talk for hours or days and never once guess that she is a download, unless she volunteered this information. Computer theorists would say that a human download passes the Turing test for intelligence and personhood. What counts for personhood is not the shape and form of a being but rather whether or not he-she-it can talk to you on a human level. If the entity can talk to you on a level such that it never occurs to you that it is not a person, then it *is* a person, whatever it looks like. The Turing test is named after Alan Turing (1912–1954), a British computer theorist who proposed it as the criterion to determine if a computer program is intelligent at the human level.

However, Turing was not the inventor of the Turing test. Jews and later Christians were the real inventors of what is now called the Turing test. For over 2,000 years, Jews and Christians have agreed that God is not a human being. God the Father has no physical form at all: He is invisible. But Jews and Christians have always insisted that God the Father is a Person. God is a Person because, and only because, He can talk to us. When Moses talked with God in the Burning Bush (Exodus 3:1–4:17), he did not see God, but he heard God's voice and concluded, correctly, that God was a Person. If God could not talk to us, He would not be considered a Person. Angels are also considered persons, precisely because they can talk to us; though in Christian theology, it is generally agreed that angels are spiritual beings.

So if they can talk to us, human downloads are people, though they will exist only as simulations in programs being run on computers. Once the technology exists to form human downloads, we can be certain that there will be humans willing to be converted into downloads. Suppose a wealthy old man is facing a slow death by cancer. Would he not reason, Why not convert myself into a download? As a download, not only will the simulation be made free from cancer, but it need never die, unless the computer running the program is destroyed. So the old man becomes a simulation. There is even no reason in principle why the original need be destroyed in the process of making the download. In which case, the download becomes a backup copy for the original human. As I write this book on my computer, I make sure that I have several backup copies. Will not humans do the same once it becomes possible to make backup copies of themselves?

Once created, the human downloads will have capabilities humans in the real world can only dream about. I have mentioned immortality, and human downloads would be far less vulnerable to bad things happening in the natural human environment. If the human download were in a computer deep underground, the download would be invulnerable to a nuclear blast overhead. Who would want to live deep underground? you might think. But the human download is not living deep underground. He is living in the simulated environment, which can be any simulated environment on the Earth. In effect, he can live anywhere on Earth—on a simulated mountaintop, on a simulated seashore, on a simulated boat in the middle of a simulated lake. He can move among these environments while never leaving the underground fortress that protects the computer wherein he has his existence. The life of a human download would be roughly like that of the "agents" in the movie *The Matrix.* Only there would be no simulated human beings to rule and exploit. Human bodies existing in the real world would be unnecessary for human downloads.

Human downloads could also speed up the rate at which their programs are run. This means that, although they would think at the usual human speed inside their simulated environment, it would appear to humans left in the real world that the downloads were thinking enormously faster than they are. A human in the real world would never be able to compete intellectually with a human download. So once human downloads are created, there will be a rush for all humans to become downloads. This rush will be encouraged by the fact that, with nanotechnology, human-to-human download would be a reversible transformation. The program corresponding to a human download could be imposed on a manufactured human body if any human download wished to return to the real world with its unavoidable dangers. Remember, anything in the real world can be simulated in the virtual reality of the downloads.

As I pointed out in the "Relativistic Spacecraft" section of the "Appendix for Scientists" in *The Physics of Immortality,* quantum computers allow human downloads to exist in miniaturized rocket payloads: a few grams of mass are sufficient to code thousands of human downloads in a simulated society, itself inside a simulated Earth environment. Real-world humans are unlikely to engage in interstellar travel. The years necessary to travel between stars would require a huge biological support system even for a single human, and many humans would be necessary to avoid psychological problems for the interstellar traveler. To expand this minimal biosphere into a diversified, Earth-like environment would require a rocket payload weighing millions of tons. A prohibitive amount of fuel would be required to accelerate such a payload to near light speed. But with only a few grams for the equivalent human download environmental system, a rocket capable of accelerating thousands of human downloads to near light speed and decelerating them at the destination star could be held in the hand of a real-world human!

In the long run—and this "long run" will probably be only a century at most—real-world humans will be replaced by virtual humans and AIs. Real-world humans will be forced to become downloads, because they cannot compete with their downloaded counterparts but also because they would be increasingly vulnerable to violence perpetrated by a small minority within the remaining real-world human community. In the present day, the main worry of government officials is nuclear terrorism. These officials actually underestimate the danger of nuclear terrorism. They mistakenly believe that, to make a nuclear bomb, a terrorist needs to obtain either highly enriched uranium or weapons-grade plutonium. This is not true. American and Russian nuclear weapons experts long ago designed a small nuclear bomb in which the thermonuclear matter, lithium deuteride, was induced to undergo fusion by a clever arrangement of chemical explosives. Fortunately, the American and Russian governments stopped research on this device, on the grounds that such a weapon would be of use only to terrorists and that anything the weapons labs developed would eventually end up on the Internet. But as technology advances, what was once a luxury affordable only by governments becomes something that can be put together in a garage by a hobbyist.

In the 1920s a tommy gun cost $200, about $4,000 in today's money. A much superior automatic weapon, an AK-47, can today be bought in the mountains of Pakistan for under $100. Any gunsmith could today make an AK-47 in a few days out of easily available materials, impossible to ban. One can today buy enough lithium deuteride to make a 100-

ton nuclear bomb for about $10, and in any case, the amount of deuterium needed for this size bomb could easily be extracted from two bathtubs full of seawater. Widely available nuclear weapons will be an inescapable feature of this century. Finally, as I pointed out in Chapter 1, the baryon-annihilation process, once it is mastered—and it must be, if the laws of physics are to remain consistent—will enable evil people to make weapons that are to nuclear weapons as nuclear weapons are to spitballs.

So real-world human history is destined to end sometime this century, either with a bang or with a download. In the longer run, over the next several centuries, the mass of the Earth and the planets in the solar system, no longer necessary as the material foundation of the biosphere, will be converted into machines to store downloads not only of humans but also of terrestrial plants and animals. As I emphasized in the first paragraph of this chapter, the biosphere on the Earth is doomed, as is the Earth as a planet. But if the Earth is taken apart long before the Sun can destroy it, not only can simulated human life and AIs survive and make use of the material composing the Earth, but simulated plants and animals can survive indefinitely as downloads also. So the true alternative is complete destruction of all life, including human life, on the Earth or survival forever as downloads for plants, animals, and humans. In view of these actual alternatives, the moral choice is clear. The Earth is destined to be taken apart so that its entire mass can be converted into (simulated) biosphere. Not only can the present-day biosphere survive as downloads, but all past biospheres can be re-created and live again, never to die.

In particular, every human being who has ever lived will one day be resurrected, never to die again. Since this is the central claim of Christianity, it is important to consider in detail why it is true. First, I shall show that it is physically possible for life in the far future to resurrect every one of us, and then show that life in the far future will in fact use this resurrection power.

Recall that the Bekenstein Bound, which I described in Chapter 2, places a definite limit on the amount of information that can be coded in any human being and in the entire universe at any given time. An upper bound to the information coded in a human being is 10^{45} bits, and an upper bound to the information coded in the entire visible universe is 10^{123} bits. Since the information content is the logarithm of the total number of possible states a human or universe could be in, if we were to exponentiate these numbers, we would have the number of distinct humans and visible universes there could possibly be. These numbers are enormous, but they are nevertheless finite. There is a finite complexity to a human being and to a visible universe at the present time.

But as the universe collapses into the Final Singularity, the amount of complexity in all the universes of the multiverse increases without limit. Not only the overall complexity but the amount of information coded in the computers of the biosphere increases without limit. Therefore, there will come a time in the far future when the amount of information required to reproduce, as a computer emulation (a simulation that is exact down to the quantum state), every human being who has ever lived is insignificant in comparison with computer capacity. Remember, computer capacity is diverging to infinity, and any finite number, no matter how large, is insignificant in comparison with infinity. Notice that life in the far future need know very little about us in order to emulate us down to the quantum state. All they will have to do is emulate all possibilities consistent with what they know about us. All they really need to know are the physical laws and the fact that the visible universe once had a radius of 10 billion light-years. Emulating all universes consistent with these two minimal facts is sufficient to guarantee that you and I will be emulated down to the quantum state.

Life in the far future would use this power to emulate us because the cost would be insignificant relative to the total resources they would have available. What counts in a cost calculation is not absolute cost but cost relative to total resources. These days, practically every American adult owns a TV. The cost of a TV relative to the income of even the poorest American is small. But fifty years ago a TV was a luxury item. Very few Americans owned one. Emulating even a single person today down to the quantum state would be impossible. The gross product of the entire world would be insufficient. But the gross product of the world today is very, very small in comparison with the arbitrarily large resources that will be available in the far future.

Since cost will be small and since life in the far future will be interested in knowing their past, of which we are a part, they will reproduce us to learn this past. In order to survive, life in the far future must drive toward total knowledge, and knowledge of the past is part of knowledge. They are also likely to resurrect us out of a sense of obligation. We are their ultimate parents, and their civilization is a descendant of ours. All their technology, including the power to re-create the past, will be built on our achievements. In comparison with their lives, our lives are miserable, just as the lives of our ancestors who lived a thousand years ago are miserable in comparison with ours. Yet all the wealth we now enjoy is built on the knowledge gained by our ancestors' suffering. Our ancestors left us this capital, mainly knowledge, which enriches our lives. We lack the power to resurrect our ancestors, so we cannot give them their due. Life in the far future will, however, have the power to bring us (and our ancestors) back into existence and share with us a tiny fraction of the wealth that without us would never have existed. We are the parents of far-future intelligent life, and "Honor thy father and mother" is a universal moral principle.

There is no difference between the human download I discussed earlier and the people resurrected by emulation in the far future, except of course the fact that the human downloads are aware of the technology being used to create them from the human originals. In both cases, simulated humans will have abilities that the original humans do not have. They will be able to modify their appearance at will inside the simulated world in which they exist. They will be able to vanish from one part of the simulation and reappear in another. Remarkably, the Gospels ascribe both of these abilities to the Risen Jesus. The Christian tradition, following Paul in 1 Corinthians 15, has been that the Risen Jesus was not merely a resuscitated corpse but instead possessed a "spiritual body," like the bodies we will all have after the Universal Resurrection.

There will indeed be a resurrection of everyone who has ever lived, and indeed we will have "spiritual bodies"—our resurrected bodies will be in the form of computer programs, which are spiritual entities, as I showed earlier in this chapter. In computer language, our future bodies will be augmented versions of our current bodies, but at a higher level of implementation. A computer program need not be run only on a physical computer; it can also be run on a "virtual computer" inside a physical computer. In fact, many of the programs now being run on desktop and laptop computers are actually being run on virtual computers simulated inside the physical computers. There is no reason to stop at a single virtual computer. A program can be run on a virtual computer being simulated on a virtual computer being simulated on a physical computer. There is no limit to the levels of virtual computer one can have. These virtual computers are the higher *levels of implementation.* Jesus had a spiritual body after his Resurrection, and in Chapter 8 I shall describe how a spiritual body could have existed at the lowest level of implementation, which is physical reality. We shall see that only God could have used the laws of physics to create a spiritual body at the lowest level of implementation. Our descendants, no matter how advanced their technology, will never be able to create a spiritual body at the lowest level of implementation.

The Three Hypostases of the Cosmological Singularity

I have discussed so far in this chapter life in only a single universe of the multiverse. The reader might wonder how the universes are fit together to make the multiverse, and what role life plays in this union. The complete multiverse of universes is pictured in Figure 3.3.

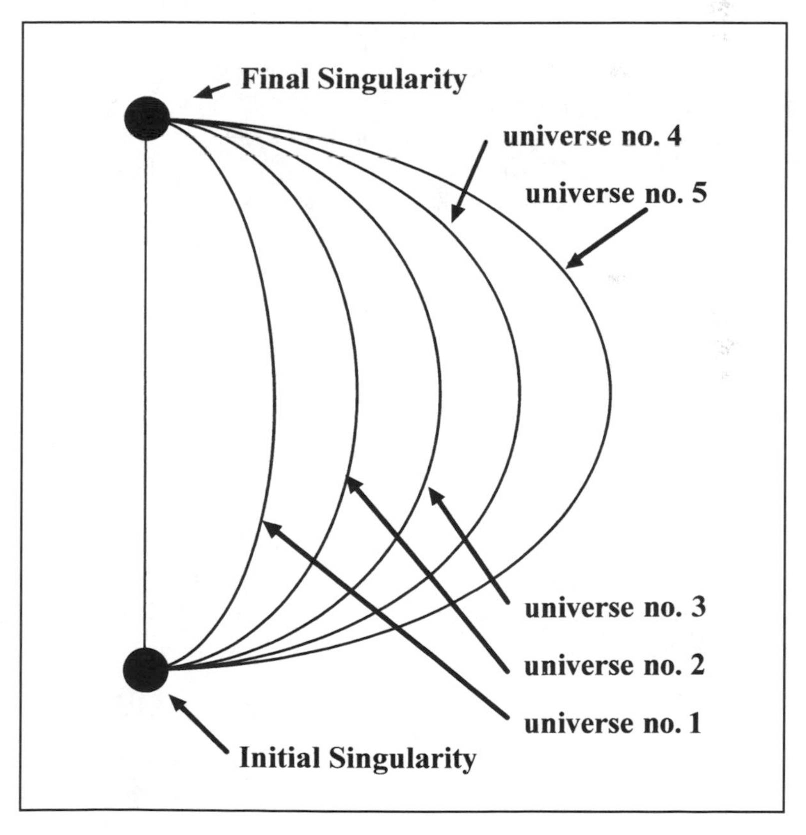

Figure 3.3. Multiverse of universes.

All the universes begin at the same initial singularity (recall that this explains why our universe is as large as it is) and end at the same final singularity. The beginning and ending of all the universes in the same singularity is necessary in order for quantum mechanics and relativity to be consistent across the multiverse. In the c-boundary construction (which is defined only inside a single universe), the final singularity is a point, the Omega Point, while the initial singularity is a 3-sphere. From the perspective of a single universe, the initial and final singularities are distinct. They are defined by the causal curves with no past end points and no future end points, respectively. Every singularity defined by a different class of curves is termed a *hypostasis* (plural, *hypostases*), the word meaning "foundation" in Greek. So a single universe has two hypostases, the initial singularity and the final singularity.

If one looks carefully at Figure 3.3, one sees that there is a third hypostasis, invisible from a single universe. It can be reached, and hence defined, only by curves that move across the multiverse, say at the same time in the various universes, in a sequence for which the universes approach zero volume at this common time. This third hypostasis connects the initial and final singularities, so they are really the same singularity, which I shall call the *Cosmological Singularity.* Since all sizes approach 0 as we approach any of the three hypostases from any direction and since these hypostases are all connected, from the multiverse spatiotemporal point of view they can be regarded as a single "point" in the usual metric meaning of this word: in this sense, the Cosmological Singularity is a single entity outside space and time. Since the Cosmological Singularity is intrinsically infinite, it is not possible to define any physical law at the Singularity itself. In other words, the Cosmological Singularity is unconstrained by physical law, not only the known physical laws but any possible physical laws.

In summary, the Cosmological Singularity is both a single unique entity without parts and, simultaneously, three distinct hypostases. In Christian theology, there is only one Entity that is like this, and it is to this Entity that we now turn.

IV

God as the Cosmological Singularity

> If anyone says that the one true God, our Creator and Lord, cannot be known with certainty by the natural light of human reason by means of the things that are made, let him be anathema.
>
> FIRST VATICAN COUNCIL, 1870

RECALL THAT IN CHAPTER 2 I POINTED OUT THAT THE UNIverse had to begin in a singularity where time and the universe began, but the singularity is nevertheless not in time or in space, nor is it material, nor is it subject to physical law. What I shall argue in this chapter is that this singularity is the First Cause of all causes. Hence, the singularity is God. I shall hereafter capitalize *Singularity.*

There are three distinct traditional proofs for God's existence: (1) the physiotheological argument (sometimes called the argument from design), (2) the cosmological argument (there must be a First Cause), and (3) the ontological argument (the existence of God is part of His essential nature). The German philosopher Immanuel Kant (1724–1804) claimed that all of these arguments had irreparable fatal defects, but his opinion arose from his ignorance of modern mathematics.

I shall now outline a version of the cosmological argument. I shall ultimately show that God is the Cosmological Singularity and that God is a Trinity. The form of the argument I shall use to demonstrate these properties will be based on the mathematical ideas used in the cosmological argument that I shall develop. So the reader is advised to become familiar with the argument and to pay close attention to the examples given to illustrate the ideas. These examples are standard in the mathematical textbooks, and they require nothing more than high school algebra as a background.

St. Thomas Aquinas (1225–1274) and Rabbi Moses Maimonides (1135–1204) defined "God" as the "First Cause" and then attempted to show that, from the existence of "causes" in the physical universe, there necessarily had to exist a First Cause, which was not in the physical universe. The First Cause was itself uncaused and the source of all causes in the universe. The verb *to create,* when applied to human action, means "to be a cause of an entity's existence." The First Cause would be the ultimate cause of all causes, so all reality would be ultimately caused by the First Cause, which would mean that the First Cause created the entire physical universe yet is uncreated Itself and outside the physical universe. These aspects of the First Cause led Aquinas and Maimonides to identify the First Cause with the Creator God of Judeo-Christianity.

There are two meanings of the word *cause,* so I shall show that the First Cause exists for both of these meanings. It is amazing that the same method can be used for both. The first meaning of *cause* is "temporal causal chain." To understand this meaning, consider the collection of causes in time that led you to read this book. Your learning about this book's existence is one. Before you could read the book, however, I had to write it, and my writing the book is another cause. You had to be born in order to read the book, and your parents getting together is another cause. But your parents in turn had to exist, and their parents, and so on. Obviously, there are a vast number of influences going back into time that resulted in your now reading this book. Is there a first such "influence"? Is there, in other words, a First Cause that is the source of all other causes and that is itself uncaused?

The answer is yes. Let us label the causes of a given event—say the reading of this book by you—by the times at which they occur. The traditional cosmological argument claimed that this temporal sequence must have a beginning, and this beginning is the First Cause. The atheists countered that there need not have been a beginning; why could not the universe be infinitely old? I shall show that even in the case of an unlimited number of causes going indefinitely far into the infinite past of an eternal universe, there still must be a First Cause.

The proof will use a trick developed by mathematicians to bring infinity into a finite distance. Imagine a straight line. A straight line has no beginning point and no end point: it stretches infinitely far in both directions. If the universe were eternal, each instant of time would correspond to a point on this straight line. The time values in such a universe, in other words, would range from minus infinity to plus infinity, with minus infinity corresponding to the infinite past and plus infinity corresponding to the infinite future. But "minus infinity" is not a point in time, since the universe has no beginning. Similarly, "plus infinity" is not a point in time, since the universe has no ending. These terms merely refer to the fact that, no matter how far back in time one goes, there is always an earlier time, and no matter how far in the future one goes, there is always a later time.

Let us now rescale the time line, that is, let us define a new time variable so that as the original time variable—call it *t*—goes from minus infinity to plus infinity, the new time variable—call it *T*—goes from some finite number to another finite number. There are many ways to do this, but a favorite among mathematicians is to use the tangent function of high school trigonometry. The tangent function is pictured in Figure 4.1.

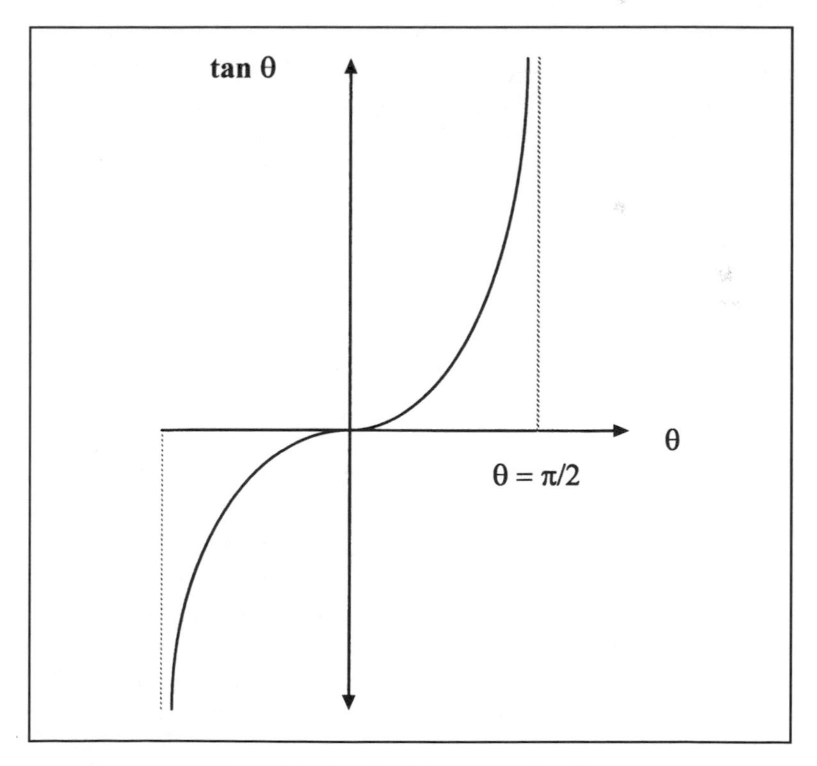

Figure 4.1. Graph of the tangent function.

Any point in the plane is defined by giving its x-coordinate and its y-coordinate, or by giving the distance from the origin of coordinates to the point and the angle that the line from the origin to the point makes with the x-axis. The distance from the origin to the point is called r, and the angle is called θ. By the Pythagorean theorem of elementary geometry, $r^2 = x^2 + y^2$, and the tangent of the angle is defined by the relation $\tan\theta = y/x$. That is, to find the tangent of the angle θ, divide the y-coordinate of the point by its x-coordinate.

Suppose we consider a series of points that are required to lie on a circle of radius 1 (this means that, although the angle θ changes, the value of r stays fixed at the value 1). Let us start when the line from the origin to the circle lies along the x-axis. In this case, we have $\tan\theta = 0/1 = 0$, and since the line and the x-axis coincide, the angle is 0. Now let us increase the value of the angle θ. This means we move the point around the circle in the counterclockwise direction. What happens to the tangent of the angle, $\tan\theta$? It increases in value, since $\tan\theta = y/x$, and y gets larger while x get smaller (remember that y started with the value 0, while x started with the value 1, since at every point on the circle of unit radius, we have $x^2 + y^2 = 1$). Now consider what happens as the angle θ approaches the right angle 90 degrees, or $\pi/2$ in radian measure. The value of $\tan\theta = y/x$ approaches plus infinity, since y is getting closer and closer to 1 while x is getting closer and closer to 0.

Now consider what happens to the tangent function as we move the point on the circle in the clockwise direction. By repeating the previous process, we see that the tangent function approaches minus infinity as the angle approaches angle -90 degrees, or $-\pi/2$ radians. In other words, the value of $\tan\theta$ varies from minus infinity to plus infinity as the angle θ varies from $-\pi/2$ to $+\pi/2$.

Let us now rescale the time variable by defining $t = \tan T$. Remember that the variable t labeled the original time variable, and this variable was the time in an eternal universe, a universe in which time varied from minus infinity to plus infinity. But the new time variable varies from $-\pi/2$ to $+\pi/2$. In this new time, T, the universe has existed for only a finite time, and it will exist for only a finite time. The eternal universe has been converted into a finite universe! But with a caveat: the "points" $-\pi/2$ and $+\pi/2$ in the new time T are not really in time, because in the original time variable, t, minus infinity and plus infinity are not time. Mathematicians say that the original time variable, t, and the new time variable, *T,* define open intervals: the end points of the *T* variable, $-\pi/2$, and $+\pi/2$, are not included, just as minus infinity and plus infinity are not included.

But in the new time variable, *T,* we can include the points $-\pi/2$ and $+\pi/2$ in time. We say we "attach" the two points to time. Then the point $-\pi/2$ is the beginning of time, and $+\pi/2$ is the end of time. Mathematicians call this procedure the *two-point compactification of the real line.* If we now return to our starting point in the argument—remember we labeled the causes by their temporal order—we see that we have established the existence of a First Cause, even if the universe were eternal. The beginning point of time, $-\pi/2$, is the First Cause, and the end point, $+\pi/2$, is the Final Cause. Notice that the point $-\pi/2$ is indeed a cause, since all causes can be traced back to it.

Granted, using the tangent function to rescale time is rather arbitrary. However, my purpose was to show that it *could* be done, rather than to show that the tangent function was the best (or only) way to rescale time. But we could pose the question of what is the "best" measure of time. The scale of time we now use is the vibration frequency of the cesium atom. But the cesium atom is just a humanly convenient way of scaling time; it is not fundamental. In fact, we know that all cesium atoms were created by nuclear fusion in the interiors of stars and in supernovas of stars, but before the first stars formed, about a billion years (measured in cesium time!) after the Big Bang, cesium did not exist.

A better scale of time duration was proposed in the nineteenth century by the great British physicists James Clerk Maxwell (1831–1879) and Lord Kelvin (1824–1907). They proposed using one of the most fundamental laws of physics, the Second Law of Thermodynamics (I always capitalize the name of this law to emphasize its central importance in physics), to define the time scale. Kelvin and Maxwell proposed to measure time by the amount of "entropy" in the universe at that time. Entropy is a physical property of materials, and it has two characteristics that make it ideal for measuring the scale of time. First, the entropy must be positive or 0—it cannot take on negative values; its smallest possible value is 0. Second, the Second Law of Thermodynamics says that the entropy of the universe must either stay the same or increase. Furthermore, the Second Law says that if a change is *irreversible*—which roughly speaking means that it cannot be undone, so that the universe cannot return to its previous state—then the entropy of the universe necessarily increases. Practically every change we witness in the actual universe is an irreversible change. So if we were to imagine going back in time, the entropy of the universe would become smaller and smaller.

But it cannot decrease without limit, since the smallest possible value of the entropy is 0. Kelvin and Maxwell proposed to label the First Cause by the 0 of entropy. If cesium time yielded a series of times in the distant past with 0 of entropy, then all these times were to be considered the same point in time. After all, if the entropy of the universe did not change even if cesium time suggested that it did, then cesium time would be physically wrong. Time is fundamentally a measure of "something happening," so if the entropy is not changing anywhere in the universe, nothing physical is changing. In this more reasonable measure of time duration, the universe began a finite time ago, at the First Cause, when the universe's entropy was 0.

The second meaning of the word *cause* is "explanation." I could explain why you are now reading this book by saying "because you bought the book yesterday." But I still need to explain why you bought the book yesterday: "because the subject matter—using physics to understand the meaning of Christianity—interested you." But now I need to explain why the subject matter interested you: "because you never could understand how it is possible for Jesus to be both man and God." But now I need to explain why you are aware of the claim that Jesus is both man and God: "because your priest said Jesus was both man and God." And so on with the list of *because*s.

In the traditional cosmological argument, it was asserted that this series of *because*s could not continue forever. There had to be a First Because, or rather, a First Cause. The atheists simply replied, "Why not?" Why can't the series go through an infinite number of levels? Let me show that there is indeed a First Cause in the explanation sense by using a modification of the mathematical technique I applied to transform infinite time into finite time.

The idea is an old one, applied most famously by Kurt Gödel (1906–1978) to prove his incompleteness theorem for arithmetic. Gödel showed that it is possible to associate a number with each mathematical statement. I shall do the same with explanation statements. Let us suppose for simplicity that each level of explanation required no more than twice the number of letters to express than the immediately preceding level. Remember that the word *level* refers to the level of causation. In the preceding example, I gave four levels of explanation, each beginning with the word *because.* (The assumption of at most doubling is not a necessary assumption—I need only to assume that at no finite level does the number of letters required become infinite—but it simplifies the argument.) Let me now write down the list of causes in order, but reducing the letter size by a factor of two at each level:

Because you bought the book yesterday

Because the subject matter—using physics to understand the meaning of Christianity—interested you

Because you never could understand how it is possible for Jesus to be both man and God

Because your priest said Jesus was both man and God

.

.

.

The dots in this rescaled list of causes represent the omitted levels of cause, which may in fact go to infinity. But since the sizes of the levels of causes decrease by a factor of two at each level, the entire list can now be given in a vertical length equal to the value of the infinite series $72 \times (1 + 1/2 + 1/4 + 1/16 + \ldots) = 72 \times 2 = 144$, since I picked the first line of explanation to be 72 points in letter size, and it is a fact of high school algebra that the infinite series $1 + 1/2 + 1/4 + 1/16 + \ldots$ equals 2.

Now since each cause in the second meaning of the word *cause* is an explanation, each set of causes—the collection of lines in the preceding list—is also an explanation. Thus the entire list is also an explanation. The *set* comprising the entire list can actually be considered equivalent to the end point of the list: *this* set *is the First Cause.* Or more precisely, the actual list on paper is a way of expressing the First Cause, just as writing down the three letters *G O D* is a way of expressing in written language the idea of the First Cause. The set of explanations—now the word *explanation* refers not to what I have written down on paper but instead to the causes in nature itself—is the First Cause. What I have shown is that if a list of explanations exists, then the First Cause exists, and this First Cause can be thought of as the set of explanations taken as a unity (this, by the way, is what the word *set* means in mathematics: all the elements or objects in the collection, but the collection considered and treated as a single object).

Returning to my earlier method of transforming the time scale using the tangent function so that the points at infinity were brought into a finite distance, it is important to note that it is actually also possible to regard the "points" at plus and minus infinity—or more precisely the points $-\pi/2$ and $+\pi/2$—as each being equivalent to the set of all points that approach each end point from inside the interval between $-\pi/2$ and $+\pi/2$. Once again, the "set of all points that approach $+\pi/2$ from inside the interval" means more than the collection of such points. It means that collection regarded as a unity. This *unity* is the point $+\pi/2$. The point $+\pi/2$ is a single point, and an infinite collection of points is an infinity of points, not a single point. But a *unity* is a single point because this is just what the word *unity* means. In mathematics, one is interested in defining an object in terms of more basic objects. One typically starts with the whole numbers, the integers 1, 2, 3, 4, and so forth. From these, one constructs the rational numbers 2/3, 4/4, and so forth, which are, as the name suggests, the ratios of integers. Then one constructs the irrational numbers ($+\pi/2$ is one irrational number, a number that cannot be expressed as the ratio of two integers, so let me take $+\pi/2$ as the irrational number to be defined) by defining the number $+\pi/2$ as the set of all rational numbers less than $+\pi/2$. As before, the word *set* means that all of these rational numbers are to be regarded as a unity, so that we get a single number, $+\pi/2$. This way of defining an irrational number is termed by mathematicians the *Dedekind Cut.*

Let us now consider another example of the First Cause, the Initial Singularity, which actually is the First Cause. One way to visualize the

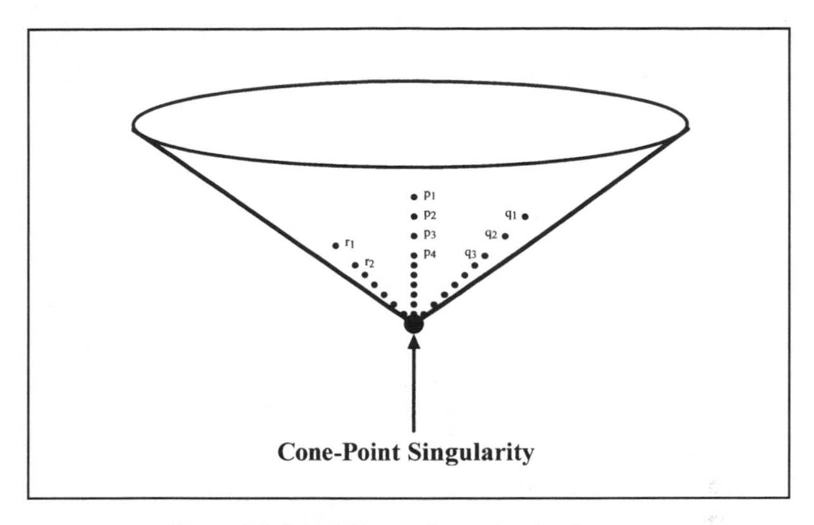

Figure 4.2. Initial Singularity as the tip of a cone.

Initial Singularity is pictured in Figure 4.2, where the universe is shown as a cone with the tip of the cone representing the First Cause.

Imagine time as the axis of the cone, with time increasing as the cone expands upward from its sharp tip. The increasing size of the cone with increasing time represents the increasing size of the universe: the universe is currently expanding—this means that the galaxies are getting farther apart as the universe ages. At present, the visible universe is a sphere some 10 billion light-years in radius, having expanded from zero size at the Initial Singularity. Also pictured in Figure 4.2 are several sequences of points on the cone that approach the tip. Notice that the closer the points of any particular sequence are to the tip of the cone, the closer they are to the corresponding points of any other sequence. We can define the "point" that is the tip of the cone as the set of all points that come arbitrarily close to the tip of the cone without actually touching that tip. That is, we pick one of the sequences shown in Figure 4.2, any sequence that consists of an infinite number of points and has points of the sequence getting arbitrarily close to the tip, and identify the entire sequence—the set of points, in other words—with the tip. There are many sequences that have this property of an infinite number of points not on the tip and whose points get arbitrarily close to the tip. But notice that the points of all such sequences get arbitrarily close to one another just as they get closer and closer to the tip. When this happens, we say that the two sequences define the same point, in this case the tip of the cone. This way of defining the infinitely sharp tip of the cone is called a *Cauchy completion.*

You might wonder, Why so much effort to define the point at the tip of the cone? Why define the tip as a set of an infinite number of points when it would be so much simpler just to take the point of the tip of the cone as itself? The reason is this: if the tip of the cone really has zero size, unlike in real cones, all of which have rounded tips if observed under a microscope, then this tip is not merely very sharp, it is *infinitely* sharp. Something that is literally infinitely sharp cannot be in space and time. It cannot be the end of an actual cone. Everything that is in space and time, every created entity, must be finite. But physics tells us that the universe began at zero size 13.4 billion years ago. So this beginning to time, where the universe was of zero size, cannot be in space and time. The beginning to time, where space and time are, so to speak, infinitely sharp, is the Initial Singularity. The word *singularity* in fact means "where physical quantities become infinite." Such an entity is outside of space and time. It is transcendent to space and time. But the method we used to define the infinitely sharp tip of the imaginary cone can be used to define the Initial Singularity in terms of points of space and time. And we can say that the Initial Singularity is a "single point" because all of the points of space get arbitrarily close to one another as they approach the Initial Singularity.

Notice that equating the set of points in a sequence approaching a singularity is essentially the same idea as the Dedekind Cut definition of an irrational number.[1] In both cases, we equate an infinite collection of points, the collection regarded as a single entity, the *set,* to another entity not in the original collection. In the Dedekind Cut, the points in the collection are all rational numbers. In the infinitely sharp tip of the cone, or the Initial Singularity, which is the beginning of time, the points in the collection are all points in space. In the Dedekind Cut, the set defines an irrational number, a number by definition not rational, a number by definition outside the collection of rational numbers. In the Initial Singularity, we have defined an entity that exists but is wholly other than space and time. Since all causal chains begin at the Initial Singularity—obviously, since the Initial Singularity is the beginning of time, though not in time—the Initial Singularity is the First Cause. The Initial Singularity is God.

It is essential to realize that, although the laws of physics require the

Initial Singularity to exist, the laws of physics cannot apply to, cannot constrain, the Initial Singularity. This is because the laws of physics are equations that are defined only for finite entities, and the Initial Singularity is an infinite entity. In fact, though it looks as if the points inside space and time are determining the structure of the Initial Singularity, this is a trick of perspective. Since we are finite creatures who reside in space and time, we necessarily study entities starting from inside space and time. So I have given a definition of the Initial Singularity using points of space and time. But the actual direction of causation should be pictured as the reverse of my definition. The Initial Singularity generates—creates—the points of space and time and the laws that govern these points and the material entities that reside in space and time. Efficient causation acts forward in time, from the First Cause at the beginning of time. The laws of physics can be considered to "flow" out of the First Cause. The laws of physics never change; they hold at all instants of time and at all points of space. But they do not apply at the Initial Singularity, because the Initial Singularity is not in space and time. It is outside of space and time, it is past the "boundary" of space and time. Before the Initial Singularity, there was nothing: no space, no time, no matter. Since there was no time either before or at the Initial Singularity, there actually was no "before." All of reality came into existence at the Initial Singularity.

If life is to guide the entire universe, it must be coextensive with the entire universe. We can say that life must have become *omnipresent* in the universe by the end of time. But the very act of guiding the universe to eliminate event horizons—an infinite number of nudges—causes the entropy and hence the complexity of the universe to increase without limit. Therefore, if life is to continue guiding the universe—which it must if the laws of physics are to remain consistent—then the knowledge of the universe possessed by life must also increase without limit, becoming both perfect and infinite at the Final Singularity. Life must become *omniscient* at the Final Singularity. The collapse of the universe will have provided available energy, which goes to infinity as the Final Singularity is approached, and this available energy will have come entirely under life's control. The rate of use of this available energy—power—will diverge to infinity as the Final Singularity is approached. In other words, life at the Final Singularity will have become *omnipotent.* The Final Singularity is not in time but outside of time. It is on the boundary of space and time, as described in detail by Stephen Hawking and George Ellis.[2] So we can say that the Final Singularity—the Omega Point—is *transcendent* to space, time, and matter.

So the laws of physics have forced us to conclude that life at the end of time—at the Final Singularity—is omnipresent, omniscient, omnipotent, and transcendent to space and time. So I identify the Final Singularity—the Ultimate Future of reality—with God (the Father). The theologian Wolfhart Pannenberg has emphasized that the Ultimate Future is what God Himself claims to be in His self-description to Moses in Exodus 3:14: "I SHALL BE WHAT I SHALL BE." God ought to know what He is. Physics is saying the same thing.

It is very important to realize that physics *can* describe the existence and properties of an entity that is not material—a singularity—and that is outside of space and time. The mathematical techniques for describing such an entity were developed by Stephen Hawking and Roger Penrose nearly forty years ago. Science is *not* restricted to describing only what happens inside the material universe, any more than science is restricted to describing events below the orbit of the Moon, as claimed by the opponents of Galileo. Like Galileo, I am convinced that the only scientific approach is to assume that the laws of terrestrial physics hold everywhere and without exception—unless and until an experiment shows that these laws have a limited range of application. The laws of physics demand that singularities exist, and this is true even in standard quantum gravity. Therefore, the laws of physics require that an entity—a singularity—exists to which the laws of physics do not apply, even though these laws predict the singularity's existence.

I show in my earlier book, *The Physics of Immortality,* that the Bekenstein Bound constrains the complexity of the universe to be finite at any time (though its complexity increases without limit as the Omega Point is approached). Therefore, the computers of the far future, which have arbitrarily large memory, can emulate down to the quantum state all those who have ever existed and allow them to live forever (in experiential time) happily in the emulated universe in the far future. This can be accomplished with a tiny fraction of the resources available to life in the far future. I showed why life in the far future would in fact resurrect us and let us live in paradise, as described in the Bible. Many people want "heaven" to lie outside of the created universe. This is the Gnostic heresy. The orthodox Christian view is that nothing exists but God and the world created by Him. Both God and His creation, as I have shown in Chapter 3, are capable of being understood in outline (though not in detail, since the Omega Point is infinite and we humans are finite).

Nevertheless, the laws of terrestrial physics show that there are worlds invisible to us (as asserted by the Nicene Creed). I refer to the other universes of the multiverse, whose existence is required by quantum mechanics. These other universes are usually considered to be a consequence of the many-worlds interpretation of quantum mechanics, but this phrase is misleading, because it suggests that there may be other interpretations of quantum mechanics. This is not so. *There is no other interpretation of quantum mechanics!* More precisely, if the other universes and the multiverse do not exist, then quantum mechanics is objectively false. This is not a question of physics. It is a question of mathematics. I gave a mathematical proof of the italicized statement in my earlier book,[3] and I outlined this proof in Chapter 2 of this book.

I am not the first to show that quantum mechanics is necessarily a many-universes theory. The first was probably the Hungarian-American mathematician John von Neumann,[4] but the great Danish physicist Niels Bohr said essentially the same thing when he claimed (the italics are Bohr's): *"However far the phenomena transcend the scope of classical physical explanation, the account of all evidence must be expressed in classical terms."*[5] In other words, Bohr, who rejected the multiverse, inferred correctly that this rejection implied that quantum mechanics does not apply at the macroscopic level. Hugh Everett, Bryce DeWitt, and David Deutsch all give alternative mathematical proofs that quantum mechanics, if correct, requires the existence of the other universes.[6] Even Roger Penrose, who does not accept the many-universes, knows perfectly well that this rejection requires him to reject quantum mechanics (at the level of the human mind).[7] If any physicists assert that quantum mechanics is correct but the other universes do not exist, then they are wrong. They have made a mathematical error, pure and simple. But maybe quantum mechanics is wrong. Maybe it is. This is a question of physics, not mathematics. However, until an experiment—and only an experiment—shows quantum mechanics to be wrong, I shall assume it to be right.

In *The Physics of Immortality,* I showed how the many universes solves the greatest of the theological problems, the problem that is the main reason people reject theism for atheism.[8] This is the Problem of Evil. In his autobiography, for example, the great evolutionary biologist Charles Darwin confessed that the existence of evil in the animal world—and the horrible suffering endured by his favorite daughter just before she died in her teens—led him to abandon his Christianity. The Problem of Evil disappears when we realize that God has maximized the good in reality by creating not just this universe but all possible universes, all of which eventually evolve into God the Father, Who is the Omega Point. I shall expand on this many-worlds solution to the Problem of Evil in Chapter 11.

But the many universes also show that the Singularity has a Trinitarian structure. I did not realize this when I wrote *The Physics of Immortality* over a decade ago, but this Trinity is in my figures and in my equations. Look at Figure VI.1 on page 184 of *The Physics of Immortality,* which is a drawing of the multiverse (reproduced here as Figure 4.3). All reality exists between the Initial Singularity and the Final Singularity.

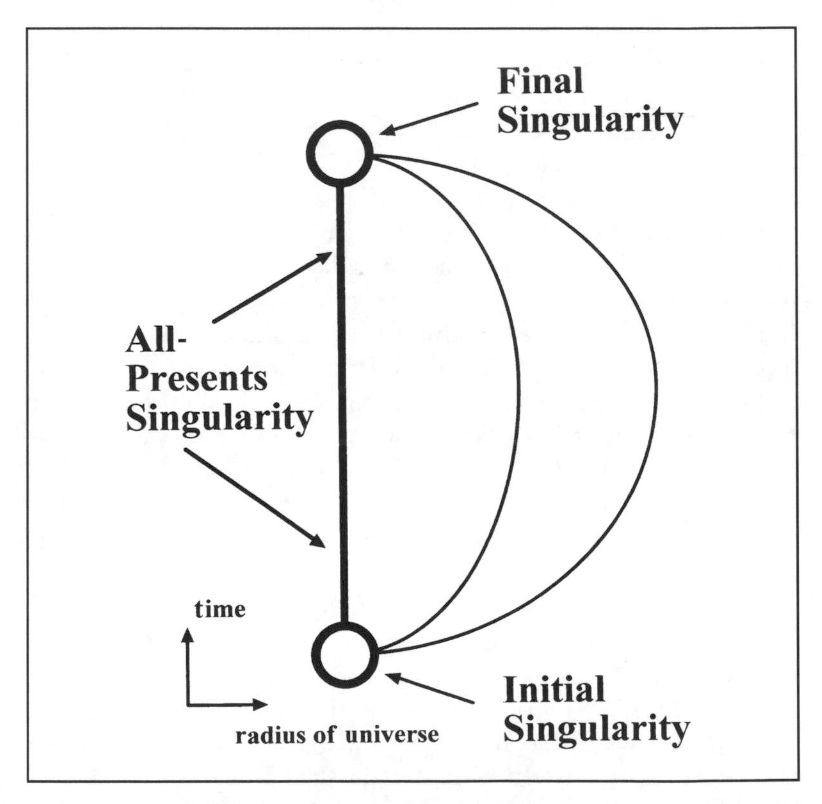

Figure 4.3. Multiverse with Initial (ultimate past), Final (ultimate future), and All-Presents Singularities. Three-Hypostatic structure pictured.

In *classical* general relativity, there is no connection between the Initial and the Final Singularities, but in *quantum* general relativity, there is a connection: the line in Figure 4.3 that connects the Initial Singularity to the Final Singularity. This is also a singularity. It exists at the "edge" of the multiverse, as indicated in Figure 4.3. It also exists at all times for all the universes in the multiverse. The quantum singularity, in other words, has a three-part structure: (1) the Initial Singularity, before which nothing existed; (2) the Final Singularity, after which nothing will exist; and (3) the singularity that connects the Ultimate Past and the Ultimate Future.

I propose to identify the Ultimate Past Singularity with the Holy Spirit (in His transcendent Godhood), on the basis of Genesis 1:2, which ends with the phrase "and the Spirit of God hovered over nothingness." This is an exact description of the Initial Singularity of the multiverse, as illustrated in Figure 4.2. I have already identified God the Father with the Ultimate Future Singularity, and I refer the reader to Wolfhart Pannenberg's extensive writings, in which he also gives reasons for thinking of God the Father as the Ultimate Future. The Son—in His Godhood, necessarily outside of time—is the connecting singularity between the Ultimate Past and the Ultimate Future. The Son is completely integrated with the Holy Spirit and God the Father. The Three are One. The Son, as is clear from Figure 4.3, was present at the beginning of the multiverse, as described in John 1:1–3: "In the beginning was the Word, and the Word was with God, and the Word was God. The same was in the beginning with God. All things were made by Him, and without Him was not anything made that was made."

The singularity is a "substance" in the same sense that electrons and protons are "substances." The key property of "substances" is that they can make their existence known by exerting effects, which can be detected. The Three Singularities—the Father, the Son, and the Holy Spirit Singularities—exert effects on space, time, and matter, even though these Singularities are outside of space and time, and are not matter. The Singularities are the divine substance, and the Son is of exactly the same substance as the Father. We have in the three parts of the Singularity—Ultimate Future, All Presents, and Ultimate Past—a full justification of the key Christian doctrine of *homoousion.* This Greek word is a compound word: *homo* means "the same," while *ousion* means "substance." So *homoousion* refers to the fact that God the Father and God the Son (Jesus in His divinity) are made of the same divine substance.

A key fourth-century debate over the nature of the Trinity was on whether God the Father and God the Son were fully equal. The orthodox side held that they were and emphasized their perspective by saying that the Father and the Son were "of the same substance" *(homoousion).* The unorthodox party believed that the Son was not fully equal to the Father and emphasized this by saying that they were of "similar substances" *(homoiousion).* This distinction was quite important, because if the Son is inferior to the Father, it is only a small step to believing that the Son is not divine at all. This is traditionally called the *Arian heresy.* (In the modern Anglophone context, it could be called the *Unitarian heresy.*) However, many at the time failed to appreciate the importance of the distinction, and considered it merely an argument over words. Their slogan was "The Empire was shaken by an argument over a diphthong." (A *diphthong* is a double-vowel sound. Thus *oi* is a diphthong, whereas *o* is a single vowel.)

A simple way to see how the Son Singularity can exert an effect in spacetime even though it is not in spacetime is to imagine a wave packet incident on the Son Singularity. As I show in the "Appendix for Scientists" in *The Physics of Immortality,* this wave packet would be reflected from the Son Singularity back into the multiverse.[9] The Son Singularity exerts an effect in spacetime—mirror reflection—even though it is not in spacetime. The boundary condition at the Son Singularity—no penetration allowed—is essentially the same as that used in electromagnetic theory to describe the reflection of an electromagnetic wave from a metal sheet. The reflection of the wave establishes the reality of the metal sheet. In March 1944, German radar pulses were reflected from metal sheets that were the skins of Allied bombers. The German radar operators inferred from this reflection that the Allied bombers were real. The bombers were indeed real, and their bombs demolished the home of a young German boy, Wolfhart Pannenberg, later to be a theologian.[10] The Son Singularity exerts the same effect on a wave function as metal exerts on a radar wave. The Son is real.

The Three parts of the Singularity are permanently distinct from one another. This fact establishes that the orthodox view of God—"God is three distinct entities of the same substance,"[11] as every Roman Catholic asserts in the Nicene Creed—is the correct one: the Trinitarian theory of the Cosmological Singularity is definitely inconsistent with the Modalist heresy.[12]

The Modalist heresy is one of the most common heresies Christians can fall into. Literally, the Modalist heretic says that the Trinity consists of just different modes of being of a single God (hence the name: *Modalist* means "mode"). But Christianity is a monotheistic religion—there is only one God—yet this God consists of three Persons. It is very natural to fall into the error of thinking of these three Persons as just different ways of looking at a single God. No: to be orthodox, you must believe that the Three are clearly distinct as Persons.

Just as the Son Singularity has a manifestation in the physical universe in the man Jesus (as I shall describe in more detail later), so the Holy Spirit has a manifestation in the physical universe as a "guiding influence." The laws of physics themselves are one expression of this guiding influence of the Holy Spirit. I have discussed this immanent aspect of the Holy Spirit in *The Physics of Immortality.* One way to visualize this is to

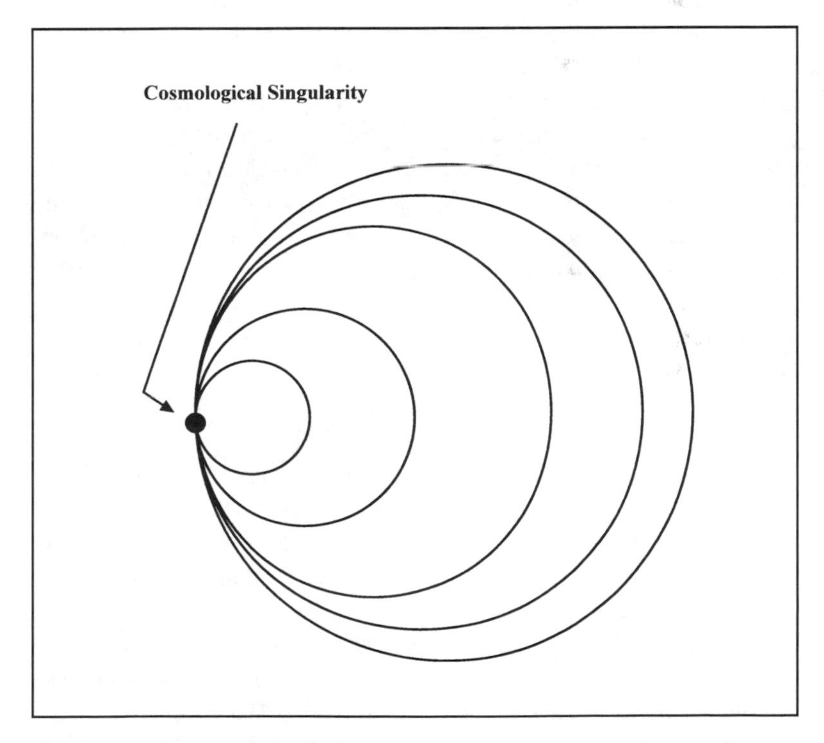

Figure 4.4. Singularity pictured in its true unity, with the multiverse forming around it. All universes of the multiverse begin at the Singularity, expand to a maximum size, and contract to the Singularity. The Singularity is shown as a point from which each universe is created and into which it ends.

regard the Holy Spirit as guiding the universes illustrated in Figure 4.4 to follow the definite trajectories pictured. The guiding influence—the Holy Spirit in His Immanence—proceeds from the Father and the Son. By contrast, the Holy Spirit in His transcendence—the Ultimate Past Singularity—is best thought of as proceeding from the Father *through* the Son. That is, the Ultimate Past Singularity arises from the Ultimate Future Singularity by means of the All-Presents Singularity.

Christology has to avoid falling into either the Modalist heresy or the Monarchianist heresy.[13] The Monarchianist heresy errs in the Arian direction, in that it claims the Son is not fully united with the man Jesus. Monarchianism also holds that the Son does not exist eternally as the Father exists eternally, so it tends to make the Son a mere creation of the Father. This aspect of Monarchianism will not concern us, since I have already established the eternal nature of the Son Singularity. The Monarchianists picture Jesus as being controlled by the Holy Spirit, or as being filled with the Holy Spirit, but not as being truly united with God. The unity of the Son with Jesus is definitely not Monarchian, as I shall discuss in more detail when I come to the theory of the Incarnation in Chapter 9. One way to see this is to note that Figure 4.4 is misleading in suggesting that our universe of the multiverse is fairly far away from the Son. If this were true, then the Son would necessarily have to be distinct from the man Jesus, who became part of our universe.

To study the distinction between the singularity of a quantum cosmology and the multiverse, it is necessary to use a technique such as the Cauchy completion technique to induce a topology on the combined multiverse and singularity. The Penrose c-boundary works only within a universe. The boundary produced by this technique is called the Schmidt b-boundary. The b-boundary has been shown to yield a topology in which the singularity is not Hausdorff separated from the points of spacetime. Roughly speaking, this means that the singularity is extremely close to every point in spacetime. Literally it means that, for any point on the singularity, it is not possible to put an open set of points between this singularity point and *any* point in spacetime proper. This "infinite nearness" shows that Figure 4.4 is misleading in suggesting that the singularity is "far away" from our universe. The b-boundary technique applied to the quantum Son Singularity pictured in Figure 4.3 also yields the fact that the Son Singularity is "infinitely close" to all points of the multiverse, just as it does in the case of the Holy Spirit and Father Singularities.

V

Miracles Do *Not* Violate Physical Law

> If anyone says that divine revelation cannot be made credible by external signs, and that therefore men should be drawn to the faith only by their personal internal experience or by private inspiration, let him be anathema.
>
> FIRST VATICAN COUNCIL, 1870

The Definition of Miracle

Ever since the eighteenth-century philosopher David Hume, most people have defined a "miracle" as a violation of physical law. But as the famous German theologian Wolfhart Pannenberg has emphasized in a recent paper, this is not the orthodox Christian definition.[1] Indeed, this definition is not implied by the biblical words for "miracle." The Greek *thaumasion*—the Latin equivalent is *miraculum*—just means "that which evokes wonder or astonishment." The Hebrew word for "miracle," *oth,* just means "sign," namely an event that indicates something other than itself. Pannenberg in his paper discusses St. Thomas Aquinas's and St. Augustine's arguments for the nonviolation of physical laws by a miracle. I would like to add to Pannenberg's discussion a mention of Aquinas's definition of *miracle* in *Summa Contra Gentiles* (Chapter 102): an event that is beyond the natural power of any creature to produce. The standard Catholic definition is due to Pope Benedict XIV (1675–1758): "a miracle is an event whose production exceeds the power of visible and corporal nature only."[2] Benedict XIV requires further that, to count as a miracle, the event must be of religious significance. Nothing is said in any of these orthodox definitions about a violation of physical law. Indeed, why should God violate His own laws? He knows what He wants to accomplish in universal history and has therefore set the laws of physics accordingly. Thus, to claim, as many modern theologians do (regrettably, even the English theologian Richard Swinburne),[3] that a miracle violates physical law is in effect to deny either God's omniscience or His omnipotence.

The claim that a miracle violates physical law also undermines the Christian doctrine of the Atonement. God could have forgiven us humans our sins without sending His Son into the universe, but doing so would have violated His laws of justice. God never, ever sets aside His laws. Therefore, His Son, the only completely sinless human being, had to suffer in our stead. God never, ever sets aside His moral laws, nor does He ever set aside His laws of physics. If we cannot trust God to keep inviolate His physical laws, then we cannot trust Him to keep His word that we will one day be resurrected to live with Him forever.

The idea that a miracle violates the laws of physics was introduced in the English-speaking world by the Deists, whose motivation was to deny the Resurrection and the Incarnation.[4] If a miracle violated physical law, if the Resurrection and the Incarnation violated physical law, then the Deists could use the strong evidence that physical laws were never violated as evidence against the Resurrection and Incarnation. Hume just continued and expanded this Deist strategy. As Pannenberg emphasizes in his paper, if we accept Hume's definition of *natural law* as a rule that is never violated, then by definition, a miracle cannot occur, and Christianity is refuted by definition.

The famous Christian apologist C. S. Lewis (1898–1963) devoted an entire book to the study of miracles and provided a defense of the orthodox position that a miracle never violates natural law.[5] Many readers, however, did not seem to understand that Lewis did defend this orthodox position on miracles. Indeed, in his article "Rejoinder to Dr. Pittenger," Lewis writes:

> I turn next to my book *Miracles* and am sorry to say that I here have to meet Dr. Pittenger's charges with straight denials. He says that this book "opens with a definition of miracles as the 'violation' of the laws of nature." He is mistaken. The passage (chapter 2) really runs: "I use the word *Miracle* to mean an interference with Nature by supernatural power" [p. 5]. If Dr. Pittenger thinks the

> difference between the true text and his misquotation merely verbal, he has misunderstood nearly the whole book. I never equated nature (the spatio-temporal system of facts and events) with the laws of nature (the patterns into which these facts and events fall). I would as soon equate an actual speech with the rules of grammar. In chapter 6 I say in so many words that no miracle either can or need break the laws of Nature; that "it is . . . inaccurate to define a miracle as something that breaks the laws of Nature" [p. 59], and that "The divine art of miracle is not an art of suspending the pattern to which events conform but of feeding new events into that pattern" [p. 60].[6]

Lewis's specific examples of miracles are unfortunate, because were they to occur as he describes them, they *would* violate physical law. He writes, "If God annihilates or creates or deflects a unit of matter He has created a new situation at that point. Immediately all Nature domiciles this new situation, makes it home in her realm, adopts all other events to it."[7] Unfortunately, annihilating or creating a unit of matter in spacetime would violate the law of conservation of mass and energy—unless the annihilating and creating were done by the direct conversion of energy into mass, a process I shall use in Chapter 8 to explain how Jesus was raised by the Holy Spirit, and to describe how a few fish and loaves of bread could be turned into enough food to feed 5,000 men. Deflecting a unit of matter would violate the law of conservation of linear momentum—unless the deflection came from the momentum carried by invisible neutrino particles. I shall show in Chapter 8 that this particular process could explain how Jesus walked on water. I doubt that Lewis had in mind these subtle physical processes, since they were not discovered until after his death.

Lewis's second example of a miracle was "If God creates a miraculous spermatozoon in the body of a virgin, it does not proceed to break any laws. The laws at once take it over."[8] But again, the creation of a spermatozoon out of nothing would violate the law of conservation of mass and energy. God could have begun a virginal conception of a male in a much cleverer way, which I shall describe in Chapter 7, and as we shall see in that chapter, the DNA evidence strongly suggests that the cleverer way was in fact the way God decided to arrange the Virgin Birth of Jesus.

Nevertheless, Lewis's heart was in the right place. Indeed, it is the influence of the supernatural—that is to say, the Cosmological Singularity, God, the only supernatural entity that really exists—acting through, not against, the physical laws of the natural world He has created, that causes miracles. I shall describe in detail in Chapter 9 on the Incarnation, which Lewis calls, correctly, the Grand Miracle, exactly how the Singularity exerts an influence in the universe of the multiverse without violating the laws that govern His creation.

Statements that miracles do not violate physical law are legion in the Christian literature. John Roach Straton, a leading self-described "fundamentalist" in the early twentieth century, denied that a "miracle" violates physical law in his famous debate with a self-described "modernist."[9] John Driscoll, a twentieth-century Roman Catholic theologian, denies that a "miracle" violates physical law in his article on miracles for the *Catholic Encyclopedia* on the Internet.[10] As St. Augustine, Lewis, Pannenberg, and a long list of Christian writers have emphasized for many centuries, for a Christian, a miracle is a very improbable event that has religious significance. Or more precisely, a miracle is an event that is very improbable from the human point of view and that can be seen to be the direct action of God—the Cosmological Singularity—in the natural world, a direct action exerted in order to make a point to us humans or to guide the universe (or individual humans) into the goal He has set.

In my description of the Omega Point theory in Chapter 3, I used past-to-future causation language, which is standard in everyday life and in most physics papers. This may have given the reader the impression that it is life that is creating the Omega Point (God) rather than the reverse. Nothing could be further from the truth. It is more accurate to say that the Omega Point, acting backward in time, via future-to-past causation, creates life and His multiverse. The quantum law of unitarity, which I used to prove the existence of the Omega Point, actually is the mathematical requirement that the two causal languages, past-to-future causation and future-to-past causation, can be translated exactly into each other. That is, quantum physics justifies teleology and, indeed, requires teleology to be true. (For a mathematical proof of this statement, see any book on quantum mechanics.)[11] Generally, however, past-to-future language will be the simpler, since the complexity of the universe, quantified by its entropy, increases with time.

There will be exceptions, however, and I will make these exceptions central to my own definition of *miracle*. I say that an event is a miracle if it is very improbable according to standard past-to-future causation from the data in our multiverse neighborhood but is seen to be inevitable from knowledge that the multiverse will evolve into the Omega Point. This definition incorporates the idea that a miracle is a very unlikely event (as in the sentence "It was a miracle he survived the car accident"), as well as Benedict XIV's requirement that a miracle have some religious significance. My definition of *miracle* thus includes both the Greek-Latin meaning and the Hebrew meaning. I further require that a miracle never, never violate any physical law.

Before a person is declared a saint of the Roman Catholic Church, a Church committee, the Congregation for the Causes of Saints, must establish that at least two miracles have occurred which can be interpreted as due to that person's intercession. Usually the miracles are cures that occurred after the potential saint was prayed to for help. The miracle is considered to be not the act of the saint but rather the act of God, Who performed it in part to give evidence that the person prayed to was in fact a holy person. In Catholic doctrine, it is God, not the saint, Who performs the miracle. The first step in being canonized—being added to the list, or canon, of saints—is beatification, an announcement that the person being investigated has passed the initial tests for sainthood. Beatification requires that at least one miracle be established as due to the potential saint's intercession. Mother Teresa has been beatified, which means the Church is satisfied that at least one miracle can be attributed to her intercession.

The agnostic Christopher Hitchens has written a book very critical of Mother Teresa, in particular arguing that the miracle attributed to her is no miracle at all.[12] Using nothing but the information provided in his book, I shall now show that, by the Catholic Church's and my definition, the event in question was indeed a miracle. Hitchens errs in using Hume's definition of a miracle, that it violated physical laws. Indeed, the event in question is not a Humean miracle, not surprising, since miracles in Hume's sense cannot exist.

According to Hitchens, the first major miracle associated with Mother Teresa occurred in 1969, when she was completely unknown to the general public. Malcolm Muggeridge, who was interviewing Mother Teresa at the time for BBC television, describes the miracle:

> This Home for the Dying is dimly lit by small windows high up in the walls, and Ken [Macmillan, the BBC photographer] was

> adamant that filming was quite impossible there. We had only one small light with us, and to get the place adequately lighted in the time at our disposal was quite impossible. It was decided that, nonetheless, Ken should have a go, but by way of insurance he took, as well, some film in an outside courtyard where some of the inmates were sitting in the sun. In the processed film, the part taken inside was bathed in a particularly beautiful soft light, whereas the part taken outside was rather dim and confused. . . . I myself am absolutely convinced that the technically unaccountable light is, in fact, the Kindly Light [Cardinal John Henry] Newman refers to in his well-known exquisite hymn. . . . This is precisely what miracles are for—to reveal the inner reality of God's outward creation. *I am personally persuaded that Ken recorded the first authentic photographic miracle* [emphasis added by Hitchens].[13]

Ken Macmillan's description of the miracle was as follows:

> During *Something Beautiful for God,* there was an episode where we were taken to a building that Mother Teresa called the House of the Dying. Peter Chafer, the director, said, "Ah well, it's very dark in here. Do you think we can get something?" And we had just taken delivery at the BBC of some new film made by Kodak, which we hadn't had time to test before we left, so I said to Peter, "Well, we may as well have a go." So we shot it. And when we got back several weeks later, a month or two later, we are sitting in the rushes theatre at Ealing Studios and eventually up came the shots of the House of the Dying. And it was *surprising* [my emphasis]. You could see every detail. And I said, "That's *amazing* [my emphasis]. That's *extraordinary* [my emphasis]." And I was going to say, you know, three cheers for Kodak. I didn't get a chance to say that though, because Malcolm, sitting in the front row, spun round and said: "It's divine light! It's Mother Teresa. You'll find that it's divine light, old boy." And three or four days later I found I was being phoned by journalists from London newspapers who were saying things like: "We hear you've just come back from India with Malcolm Muggeridge and you were the witness of a miracle."[14]

It is obvious that both Hitchens and Macmillan believe a miracle must involve a violation of physical law. It is equally obvious that no violation of physical law occurred in this miracle. For miracle it was, but the miracle was the double *coincidence* that the new Kodak film was made available just in time for the Mother Teresa interview, and it was used for just the House of the Dying shot. The film quality was—in Macmillan's words—"surprising," "amazing," and "extraordinary." The film quality was a *wonder,* and "wonder" is just what the word *miracle* means. The effect of this new film was to make Mother Teresa into a worldwide star, and this in turn enabled her to preach the Gospel in India, an act the Indian government opposes. Hitchens makes clear in his book that it was Mother Teresa's star power which made the Indian government reluctant to shut down her operation. So we have an improbable event whose result is to allow the preaching of the Gospel. "An improbable event whose effect is to carry out God's plan for the universe" is a more exact expression of what Christians mean by *miracle.*

Mother Teresa herself always emphasized that a miracle is the action of Providence—which means God acting through natural law, pushing the universe in the direction He wishes it to go. Hitchens quotes another example:

> One day Sister Frances, from the city of Agra, phoned Mother Teresa asking for urgent help.
>
> "Mother, I need 50,000 rupees. Over here there is a crying and urgent need to start a house for the children."
>
> Mother Teresa replied: "That is too much, my daughter, I will call you back; for the moment we have nothing." . . . A short time later the phone rang again. It was a press agency. "Mother Teresa? This is the editor of the agency. The Philippine government has just awarded you the Magsaysay Prize. Heartfelt compliments! It involves a considerable sum."
>
> Mother Teresa: "Thanks for letting me know."
>
> The editor: "What do you plan on doing with the 50,000 rupees from the prize?"
>
> Mother Teresa: "What did you say? 50,000 rupees? I think the Lord wants us to build a home for children in Agra."[15]

The Philippine government did not violate physical law when it awarded Mother Teresa 50,000 rupees. Mother Teresa once again correctly interpreted as a miracle the coincidence that, just after Sister Frances requested 50,000 rupees of her, she was informed she had been awarded exactly the amount requested. So, using only the evidence of Hitchens, Mother Teresa has two genuine miracles to her credit, enough to qualify for sainthood.

Another example is the Miracle of the Sun at Fatima. On October 13, 1917, a large crowd, estimated between 10,000 and 80,000 people, gathered in a field outside the small village of Fatima in Portugal because three small children had announced that there would be a miracle that day.[16] The children said that they had seen an apparition of the Virgin Mary in that field once a month for several months, and that the Virgin had told them to come back at this particular time. She had promised the children she would provide evidence—a miracle—that she was indeed present, although none but the children could see her. The Monday, October 15, evening edition of a leading Lisbon newspaper, *O Seculo,* carried the headline "How the Sun Danced at Midday in Fatima."[17] A photograph of the crowd at Fatima observing the solar phenomenon, taken by a photographer from *O Seculo* on October 13, appeared with the article.[18] There are no photographs of the solar phenomenon seen by the people at Fatima. This article, which appeared in a secular, not a Catholic, newspaper, began a huge controversy in Portugal. The Sun was seen to move in an unusual way at noon at Fatima, "danced" in the words of the headline, and this motion was seen by thousands. What actually happened?

The solar motion was localized at Fatima. No one in Lisbon, no astronomer anywhere, saw the Sun dance at noon on October 13, 1917. Some people in the crowd believed they saw the Sun fall out of the sky. Stanley Jaki has studied all available reports of eyewitnesses, and his best guess of what was seen is that the Sun rotated.[19] That is, the Sun appeared to spin in its position in the sky. There was also a haze covering the Sun at noon, which was why people were able to look at it. The German meteorologist K. J. Stöckl has pointed out that, when the eye looks at the Sun directly, just before the light level becomes uncomfortable, the Sun appears to spin.[20] This effect is definitely real.[21] Conditions at Fatima were ideal for this optical illusion to occur, so optical illusion is the most probable explanation of the Miracle of the Sun.[22]

The movement of the Sun at Fatima would be a phenomenon of the human retina, not the Sun. No natural laws were violated in the Sun at Fatima. Nevertheless, it was a miracle, for two reasons. First, as Jaki recounts, it was an announced miracle. The announcement that a miracle would occur that day is why at least 10,000 people were at Fatima on

October 13. Second, Christianity was restored in Portugal by this miracle. In 1917 militant atheists formed the government of that country, and they wished to suppress the Catholic Church. A local government official, an atheist who believed the children were making up their story of seeing the Virgin Mary, had arrested the three children in September. After the Miracle of the Sun, the suppression of the Catholic Church was politically impossible.

The Gnostic Heresy and the History of Science

The idea that miracles violate physical law is actually a form of the Gnostic heresy, not Christianity.

The word *heresy* comes from the Greek *hairesis,* which means "choice," the implication being that one chooses the heresy, rather than it being forced on one, in the way that logic and experimental evidence force one to accept the laws of physics. I accept the laws of physics, in particular quantum mechanics and relativity, which is why I accept not only the existence of God but His Trinitarian nature. I have no choice in accepting the Trinity if I wish to follow where the laws of physics lead. Most physicists choose to abandon the laws of physics when they realize that these laws are leading to God. Most physicists, in other words, are heretics, not so much to Christianity but to science.

There are really only two great heresies to Christianity: the Arian heresy and the Gnostic heresy. The Arian heresy denies the full divinity of Jesus, and we shall discuss this heresy in more detail in Chapter 9. The Gnostic heresy is connected with the proper definition of *miracle,* so it will be discussed here. The Problem of Evil, which we shall resolve in Chapter 11, is the fundamental cause of the Gnostic heresy. The Problem of Evil is simply, Why is there evil at all? If God is all-powerful, all-knowing, and all-good, then why does He allow evil to exist? The Gnostics answer this problem by denying that God is all-powerful. Instead, they say, there are two gods, one good and one evil, who contend for power. This divine dualism is manifested in a further dualism, between matter and spirit. According to the Gnostics, the evil god created the material world. The spiritual world is the creation and domain of the good god. Our souls, being spiritual, are the creation of the good god and yearn to return to the spirit world to be with this good god. Unfortunately, our souls have been imprisoned in our material bodies by the evil god, and thus are subject to pain inflicted by the evil in this evil material world.

The Gnostic heresy has arisen many times in the 2,000 years of the Christian era, and thus it has many names. It was first called the Marcionite heresy, after the Christian bishop Marcion, who was expelled from the Church for advancing this heresy in 144. Marcion argued that the picture of the God of creation, as conveyed in the Old Testament, was quite different from the loving God of the New Testament. The Old Testament God was a God of war, slaughtering people right and left.

Consider God's words to Moses just before He parts the Red Sea: "But lift thou up thy rod, and stretch out thine hand over the sea, and divide it: and the children of Israel shall go on dry ground through the midst of the sea. And I, behold, I will harden the hearts of the Egyptians, and they shall follow them: and I will get me honor upon Pharaoh, and upon all his host, upon his chariots, and upon his horsemen. And the Egyptians shall know that I am the LORD, when I have gotten me honor upon Pharaoh, upon his chariots, and upon his horsemen" (Exodus 14:16–18). Why is an all-good and all-loving God hardening the hearts of his children, encouraging them to rush to their deaths? Who but a god of war would want to win honor by destroying an army? Or consider the order of God as recorded in Numbers 31: "And they warred against the Midianites, as the Lord commanded Moses; and they slew all the males" (verse 7). "Now therefore kill every male among the little ones, and kill every woman that hath known man by lying with a male. But all the women children that hath not known a man by lying with him, keep alive for yourselves" (verses 17–18). How can an all-good God give such a monstrous order? How could an all-good God permit what the Israelites did after they took the city of Jericho: "And they utterly destroyed all that was in the city, both man and woman, young and old, and ox, and sheep, and ass, with the edge of the sword" (Joshua 6:21).

There are many passages like these in the Old Testament, and Marcion concluded that the god who gave such orders was innately evil. Since, according to Genesis, this god also created the material universe, he must also be evil. Marcion believed that the Old Testament was the document of this evil god, so he proposed eliminating the entire Old Testament, and much of the New, from the Christian canon. An implication of the Marcionite heresy is that the Jews, who carried out the evil orders described in the Old Testament, are servants of the evil god and are thus themselves innately evil. The Marcionite heretics were expelled from the Church by the end of the fourth century. The sociologist of religion Rodney Stark has suggested that a majority of Christians in the first three centuries of the Church's existence were converted Jews, and that the Jews protected Christianity from the Marcionite heresy. The Christian Jews would naturally have been dubious of the claim that they were innately evil.

In the fourth century, the dualism of Gnosticism reappeared in the form of Manichaeanism, originally of Persian, not Christian, origin. It was strongly opposed by the state, since Christianity had become the established religion of the Roman Empire in the early fourth century, and disappeared from western Europe by the end of the fifth century and from the eastern empire by the end of the sixth. Gnosticism appeared again in southern France in 1020, called then the Albigensian heresy, and was not suppressed until the fourteenth century. The Holy Inquisition was created in 1231 mainly to act against the Albigensians. St. Dominic founded the Dominican Order in 1215 in order to oppose intellectually the doctrines of the Albigensians, and the Inquisition was largely under the control of the Dominicans until it was abolished (or rather, renamed the Congregation for the Doctrine of the Faith) in the twentieth century.

The key feature of all these versions of Gnosticism is the dualism of two gods, one good and the master of a spiritual universe, the other evil and the creator of the material universe. According to the Gnostics, we humans are kept in ignorance of the spiritual world. We become aware of the spiritual world only through the invasion of the evil material world by denizens of the spirit world. These beings use their power of good to act in the material world, and these acts are what we call "miracles." In the Gnostic worldview, miracles really are violations of the laws of physics, since the physical laws are the laws that govern the evil material world. The Gnostics are uninterested in studying the laws or nature, or even in establishing if such laws exist, because the material world itself is of no importance to them. We are prisoners in this material world, and what is important is escaping from it and learning about our true spiritual nature and about the nature of the spiritual world created by the good god. This knowledge, for the Gnostics, is the only true knowledge, and this secret knowledge of spiritual reality gives them their name: *gnosis* is Greek for "knowledge."

St. Augustine, who was briefly a Manichaean himself, in his book titled *Against the Manichaeans*, makes clear why he wished to emphasize that a miracle does not violate the laws of nature: the material world in the Christian worldview was the creation of God, who knew what He was doing. God never has to act contrary to His own creation. To suggest that He does act contrary to His own creation is to suggest that it is not really His creation but the creation of another god of equal power, and thus we are forced into the Gnostic worldview. No! Spiritual reality and material reality are equally creations of the One God, and totally subject to Him, Who is unchanging. His Will is forever constant and dependable. His laws never change, just as His Will never changes. Furthermore, since His laws are His direct creation, studying His natural laws is as pious an act as studying the Bible.

If God can change His mind about His law, then salvation from Jesus' death on the Cross is at risk. God may change, without informing us, the rules about what is necessary to obtain salvation. On the contrary, God's law is eternal and never changing. Some Christians may wish to believe that there is a distinction between moral law and natural law, but the Church has always held that there is no distinction, and in fact it has always attempted to derive moral law from natural law. I shall argue in Chapter 7, where I discuss the Immaculate Conception, that there is indeed no ultimate distinction between moral and natural law. All moral judgments are really judgments about matters of fact. The value-fact distinction does not exist.

Related to the Gnostic heresy is the claim that there cannot be any laws of physics, because even saying that there are laws limits God's power to do as He wills. In practice, this is the same as saying that God can set aside the laws of physics whenever He wishes, and that to say He never will set aside the physical laws is to limit God's power. Pope Urban VIII, in a private conversation with Galileo, used precisely this argument in his rejection of the Copernican system. As Galileo summarized this argument in his *Dialogue on the Great World Systems* (the argument was presented by "Simplicio," the defender of the geocentric worldview):

> I know that both of you, being asked whether God, by His infinite power and wisdom, might [generate effects by a means other than in your theory] that you would answer that He could, and also that [He] knew how to bring it about in many ways, and some of them above the reach of our intellect. Upon which I forthwith conclude

> that, this being granted, it would be an extravagant boldness for anyone to go about to limit and confine Divine power and wisdom to some one particular conjecture of his own.[23]

Urban VIII was speaking off the cuff, and indeed, his argument was heretical (as he himself later admitted). This argument would undermine the orthodox Christian view that nature is the rational creation of a rational God and, incidentally, make the scientific study of nature impossible. Earlier in the *Dialogue,* Galileo had refuted this argument:

> Surely, God could have caused birds to fly with their bones made of solid gold, with their veins full of quicksilver, with their flesh heavier than lead, and with wings exceedingly small. He did not, and that ought to show something.[24]

There is a passage in the Qur'an (6:64) that has been interpreted by most traditional Muslims to mean that there cannot be any laws of physics because having unchangeable laws would limit God:

> The Jews have said, "God's hand is fettered." Fettered are their hands, and they are cursed for what they have said. Nay, but His hands are outspread; He expends how He will.[25]

The word *fettered* can also be translated as "chained."[26] In other words, if there exist laws of physics that are never altered, then God would be constrained by the very existence of these laws. Instead, the Will of God must be entirely unconstrained, and He must be viewed as free to change the laws of physics from moment to moment. Further, there is a curse on the head of anyone who dares to claim that the laws of physics are fixed and unchanging. Such a worldview does not encourage the search for unchanging physical laws.

In fact, it actively discourages the very idea of physical laws. In 1982, the Institute for Policy Studies in Islamabad, Pakistan, recommended that science textbooks be modified to emphasize that all change was due not to the action of physical law but to God:

> There is latent poison present in the subheading *Energy Causes Changes* because it gives the impression that energy is the true cause rather than Allah. Similarly it is unIslamic to teach that mix-

> ing hydrogen and oxygen automatically produces water. The Islamic way is this: when atoms of hydrogen approach atoms of oxygen, then *by the Will of God* water is produced.[27]

The implication being that God may change His mind in the next instant, and water would not be produced. The Muslim theologian Abu Hamid Mohammed al-Ghazali (1058–1111), famous for making Sufism (Muslim mysticism) part of orthodox Islam, wrote a book, *The Inconsistency of the Philosophers,* attacking the idea of cause and effect, and hence arguing that scientific knowledge is impossible. Rather than follow natural philosophers (scientists) and say that fire burns cotton:

> This we deny, saying: the agent of the burning is God, through His creating the black in the cotton and the disconnection of its parts, and it is God Who made the cotton burn and made it ashes either through the intermediation of the angels or without intermediation. For fire is a dead body, which has no action, and what is the proof that it is the agent? Indeed the philosophers [scientists] have no other proof than the observation of the occurrence of the burning, when there is contact with the fire, but observation proves only a simultaneity, not a causation, and, in reality, there is no cause but God.[28]

Sufi theologians followed al-Ghazali and insisted that physical laws did not exist because God destroys and re-creates the universe from one instant to the next.[29] In my own rather extensive studies in Islam, I have never been able to find a single significant scientific discovery made in the entire history of Islamic civilization up to the twentieth century. The examples in the literature of Islamic scientific achievements are essentially trivial. All modern physics and astronomy descends from the work of the Christians Galileo (1564–1642) and Copernicus (1473–1543), who effectively ignored the "work" of Islamic "scientists" and instead started with the work of the Greeks Archimedes (290–211 B.C.) and Ptolemy (A.D. 100–170), respectively. From the point of view of science, Islamic civilization did not exist. I attribute this fact to the Islamic theological doctrines against the idea of experimentally confirmed natural law just quoted, combined with the fact that, throughout Islamic history, anyone disagreeing with the prevailing theology has been regarded as an apostate, and the overwhelming number of Islamic jurists have agreed: the penalty for apostasy is death. No one is going to search for the laws of nature if even suggesting they exist makes him or her subject to the death penalty. A conference of seventeen Arab university presidents was held in Kuwait in 1983. The major topic of discussion was "Is science Islamic?" The Saudi delegation argued that science is not, being intrinsically secular and, hence, automatically against Islamic beliefs.[30]

There is a (false) tradition, possibly originating with Christian critics of Islam, that when Muslim armies took the Egyptian capital city of Alexandria, the head of the Muslims, the second Caliph Omar ('Umar ibn al-Khattab, 586–644), ordered that the books in the library be burned to heat the bathwater of the Muslim soldiers. If the books disagreed with the Qur'an, they were heretical, and if they agreed with the Qur'an, they were superfluous. In either case, they should be destroyed. In reality, the Great Library of Alexandria ceased to be mentioned by eyewitnesses after about 100 B.C., and there are no entries in the list of head librarians after that time, so probably the library ceased to exist by 100 B.C.,[31] possibly destroyed in the chaotic reign of the Egyptian king Ptolemy VIII, known to history as Ptolemy the Psychotic. (I'm not kidding, this really was the title given to him by Greek historians after his death; *psychon* was the word they used; "hostile" is another possible translation.) So neither the Christians (as has often been charged) nor the Muslims were responsible for the destruction of the Great Library. The claim that religious fanatics burned the library is a myth. But there was a crucial difference between the Christian and Muslim responses to this myth. Christians felt the need to apologize; many Muslim scholars, believing the myth, quoted it with approval. Indeed, books disagreeing with the Qur'an should be destroyed, and there is no need to read any book outside the Qur'an.

There is one exception to the rule that there were and are no significant Muslim scientists: Mohammed Abdus Salam (1926–1996). Salam was one of the main creators of the Standard Model of particle physics, a theory that is absolutely central to this book. I described the Standard Model in Chapter 2, and as we shall see in Chapter 8, it is crucial in understanding how the Resurrection of Jesus was accomplished. Salam deservedly received the Nobel Prize in physics in 1979 for his work on the Standard Model, and his idea that quantum gravity can make quantum field theory finite is crucial to the Omega Point theory of the Ultimate Future, though too technically complex to describe here.[32] Salam was a Muslim in the sense that he called himself a Muslim, and all who knew him are convinced that he was completely sincere in thinking himself a Muslim.

Salam is the exception who proves the rule. By an act of the Pakistani Parliament in 1974, the Ahmadi sect of Islam, to which Salam belonged, was declared heretical and subject to persecution.[33] Salam's coauthor of *Islam and Science,* Pervez Hoodbhoy, reported on his website in 2002: "My next-door neighbor, an Ahmadi, was shot in the neck and heart and died in my car as I drove him to the hospital. His only fault was to have been born in the wrong sect."[34] Salam himself left his native Pakistan in the 1950s, realizing that, in that country, doing serious physics would be impossible. Had Salam remained in Pakistan and nevertheless achieved what he eventually achieved as a physics professor at the University of London, he would have become the most prominent Ahmadi in Pakistan, and as such, he would probably have met the fate of Hoodbhoy's neighbor.

Muzaffar Iqbal, in a book also entitled *Islam and Science,* does not once mention the greatest Islamic scientist of all time, Abdus Salam, even though the book was written in 2002, largely to counter the book by Hoodbhoy and Salam, and claimed to be a detailed examination of the scientific achievements of Islam. Iqbal mentions only Hoodbhoy. Salam is a heretic, and hence not a Muslim. In their book, Hoodbhoy and Salam show that virtually all Muslim scientists now regarded as significant were persecuted in their own times. Like contemporary defenders of Islam, the French physicist and Roman Catholic Pierre Duhem (1861–1916) tried to prove that the Christian scholars of the Middle Ages made important contributions to physics, for example, by introducing the concept of inertia. Neither the Muslim scholars of the so-called Golden Age of Islam (roughly 700–1100) nor the medieval Christian scholars made any significant contribution to physics. As I pointed out earlier, neither Copernicus nor Galileo was aware of these "significant contributions."

Nevertheless, modern science was a creation of Christian civilization. The creative period of Greek physics and astronomy ended about 100 B.C. This end date is important, because it is occasionally claimed that it was the rise of Christianity that ended Greek science. Not so, as the following list of Greek physicists and astronomers and their dates show:

Pythagoras of Samos (580–500 B.C.), the first great Greek mathematican. His school discovered the theorem bearing his name and established the existence of irrational numbers.

Socrates (470–399 B.C.)

Plato (428–347 B.C.), the philosopher who believed all physics should be based on mathematics

Theaetetus of Athens (417–369 B.C.)

Eudoxus of Cnidus (395–337 B.C.)

Aristotle (384–322 B.C.), the philosopher who argued that motion cannot be described by mathematics. Galileo's main opponents were followers of Aristotle.

Euclid of Alexandria (f. 323–285 B.C.)

Aristarchus of Samos (f. 310–230 B.C.), the first to propose a heliocentric solar system

Archimedes of Syracuse (290–211 B.C.)

Apollonius of Perga (260–190 B.C.)

Hipparchus of Nicaea (200–127 B.C.)

Hypsicles of Alexandria (190–120 B.C.)

End of Greek science's creative period (c. 100 B.C.)
(End of the Great Library at Alexandria)

Hero of Alexandria (f. A.D. 60)

Ptolemy of Alexandria (A.D. 100–170)

Diophantus of Alexandria (f. A.D. 250)

Pappus of Alexandria (f. A.D. 320)

Hypatia of Alexandria (A.D. 370–415), killed by a Christian mob

These dates indicate that the listed mathematicians and physicists—in Greek times, there was no difference—overlapped with and could have known one another. In some cases, we know that they did know one another and were related teacher to pupil, as I was the postdoctoral student of John A. Wheeler (the man who named the black hole and whose most famous student was Richard Feynman). Wheeler was in turn the postdoc of Niels Bohr, who was the postdoc of Ernest Rutherford, who discovered the atomic nucleus, and J. J. Thomson, who discovered the electron. By 100 B.C., the overlap ceased, and Greek science with it. One of my teachers when I was an undergraduate, the historian of science Giorgio de Santillana, the greatest Galileo scholar of his generation, has given several reasons for the fall of Greek science (he gives 200 B.C. as the ending date).[35] Chiefly, the reasons were (1) the bureaucratization of science with the rise of the Hellenistic empires, begun by Alexander the Great and ending with the Roman Empire, and (2) the rise of the Gnostic mystery cults, which undermined the idea that the material order was a road to ultimate knowledge.

There were no intellectual barriers preventing modern science from beginning development in 100 B.C. Basically, all Copernicus did in 1543 was restate the geocentric universe of Ptolemy in a heliocentric frame of reference. (This was not as easy as it sounds; it took a first-rate mathematician to do it. But mathematical genius the Greeks had, and all the ideas Ptolemy used had been developed by 100 B.C. Ptolemy was a textbook writer, not an original mathematical astronomer.) Aristarchus of Samos had written a book (now lost) by about 300 B.C. describing mathematically a heliocentric solar system. Copernicus even used Ptolemy's observational data, data available long before 100 B.C.

The Dutch historian of science H. Floris Cohen has given a particularly striking example of how easy it should have been for the Greeks to begin modern science in 100 B.C.[36]

The Greeks knew well before 400 B.C. that if a vibrating string's length were cut in half, the tone would be raised an octave. The fifth corresponded to a 2:3 length ratio, and so forth. Also long before 100 B.C., the Greeks had two theories of sound, one that it is a vibration of the air (the correct explanation), and the other that it consists of some sort of particle transfer. But it was not until A.D. 1563 that the Italian Giovanni Battista Benedetti (1530–1590) developed in a mere forty-line paragraph the modern theory that the wavelength of the sound wave equals the length of the string, making for the first time a quantitative connection between the ancient theory of sound and the ancient theory of musical sound. The Greeks had all the necessary ideas, but they never made the connection, obvious as it seems to us.

If bureaucratization of science and the growth of Gnosticism are the reasons for the end of Greek scientific development, then our own civilization is gravely at risk. At the end of the nineteenth century, there began an interest in the occult, a trend that has been steadily increasing in Western civilization to the present day. A manifestation of this wide interest is the huge worldwide success of the Harry Potter fantasy novels. These novels develop the full implications of the Gnostic dualist worldview: there is the mundane world of physics, inhabited by ordinary people subject to these laws, and a hidden magical world, inhabited by wizards and witches who are capable of manipulating the more powerful, and spiritual, power of magic. Harry Potter himself is a young wizard who is attending Hogwarts, a school for the training of wizards and witches. Ordinary people, given the derogatory name "Muggles" by the wizard community, are unaware of the magical world. Worse, the magicians always defeat the Muggles when the former come into conflict with the latter. Not surprisingly, the wizards—even the "good" wizards such as Harry Potter and his mentor, Albus Dumbledore, the headmaster of Hogwarts—treat ordinary, nonmagical people as inferiors whose feelings need not be considered. The Gnostic leaders held a similar view of ordinary people.[37] The medieval Gnostics called themselves the *Cathars* (Greek for the "pure"), and their leaders were called the *Perfecti.*[38] Needless to say, neither the wizards in the Harry Potter fantasy nor any of the Gnostics who have arisen over the past twenty centuries have expressed an interest in learning the science of the natural world. There is no course in physics or chemistry at Hogwarts, and the only astronomy course is devoted to astrology. Interest in magic drives out interest in natural science.

The early Roman Catholic Church, following the lead of St. Augustine, opposed witchcraft and magic, not because it was the work of the Devil but because it did not exist! St. Boniface (675–754), the Wessex Saxon who began the conversion of Germany to Christianity, wrote that it was "unchristian" to believe in witches and werewolves.[39] The emperor Charlemagne (742–814) in 785 imposed the death penalty on anyone who burned witches at the stake, on the grounds that such burning was a "pagan custom."[40] In 820, St. Agobard, Bishop of Lyon (769–840), claimed the idea that wizards could cause bad weather was nonsense. Catholic disbelief in witches was codified for centuries thereafter as official Church law in *Canon Episcopi,* which stated that claims of broomstick flying and human-animal transformation were hallucinations, and whoever believed in them was "beyond doubt an infidel and a pagan."[41] Coloman (1070–1116), king of Hungary from 1095, refused to establish laws against witches, "since they do not exist."[42] John of Salisbury (1115–1180), the secretary of St. Thomas à Becket, the archbishop of Canterbury who was assassinated in his cathedral by supporters of Henry II, contended that the idea of a witches' Sabbath was a fable.[43]

Alas, Roman Catholic disbelief in the power of witchcraft was not to last. In 1484, Pope Innocent VIII (1432–1492) issued the bull *Summis Desiderantes Affectibus,* wherein he denounced the increase of witchcraft in Germany and authorized the Dominican inquisitors Heinrich Institor and Jakob Sprenger (who happened to be his sons) to suppress it.[44] Two years later, Institor and Sprenger published the first great encyclopedia of witchcraft, *Malleus Maleficarum,* meaning "The Hammer of Witches." The Church had completely reversed itself, for the subtitle of *The Hammer* was "to disbelieve in witchcraft is the greatest of heresies."[45] What had caused this radical change between the twelfth and fifteenth centuries?

For one thing, the Great Plague, or Black Death, which between 1347 and 1351 killed about a third of the European population, inspired a huge rise in the belief that demonic powers were active in the world. However, natural disasters, even on the scale of the Great Plague, would not have led to the belief that demons, acting through witches, caused the catastrophes unless the intellectual case had already been made for the belief. This case had been made by the Dominican friar St. Thomas Aquinas (1225–1274) in his greatest work, *Summa Theologica.* Aquinas had based his theology, including his theory of miracles, on Aristotelian physics, which, as we have seen, did not allow crucial Christian miracles such as the Virgin Birth and Jesus' Resurrection. So Aquinas modified the standard Augustinian view of miracles. Miracles for Aquinas involved God overcoming Aristotelian law. If the supernatural power of God set aside the normal course of nature, then the natural course of nature could also be set aside by the action of demons, who could be invoked by people: witches and wizards. The Old Testament commandment "Thou shalt not suffer a witch to live" (Exodus 22:18),[46] earlier interpreted as imposed because believing oneself to have magical powers from the Devil was equivalent to believing the Devil equal to God in power (i.e., believing in the Gnostic heresy), was now interpreted as a commandment to destroy the vessel of the Devil's power.

The alert reader will have noticed that the Dominican Order was founded to combat the Gnostic heresy in its medieval form, yet in the end, it was the Dominican Order that played a major role in persuading the Catholic Church to accept the existence of witches, in effect accepting Satan's power as equal to God's, which is the essence of the Gnostic heresy. In the end the Dominican Order came to defend what it had been created to fight. The historian H. R. Trevor-Roper opines that this evolution was due in part to the close association with the Gnostic heretics.[47] In hearing repeated confessions (under torture) by wizards and witches, the inquisitors started to believe that where there's smoke, there's also fire. But I think the powerful pull of Gnostic philosophy also was a significant effect. The heresy must have deep roots in the human psyche or it would never have reappeared again and again in history. In the absence of a convincing solution to the Problem of Evil, it is only too easy to believe in an evil god of equal power to the good God.

Fortunately, the old belief that magic did not exist, that natural law was the unchanging Word of God, had sufficient inertia to inspire the first scientists, Copernicus (1473–1543) and Galileo (1564–1642). Note that Copernicus was ten years old when Innocent VIII promulgated his witchcraft bull. The idea that a personal God has decreed the unchanging laws of nature is unique to Judaism and Christianity, and there is strong evidence that this is why modern science arose in the Christian West. Stanley L. Jaki and Rodney Stark have written books arguing that Christianity and its idea of unchanging natural law arising from an unchanging God was essential for the development of modern science.[48] Jaki is a Catholic priest, and Stark is an Evangelical, so one might be tempted to suspect a bias toward Christianity in these scholars. However, the Chinese Academy of Social Sciences of the People's Republic of China came to a similar conclusion in 2002:

> One of the things we [the Chinese Academy] were asked to look into was what accounted for the success, in fact, the pre-eminence of the West all over the world. We studied everything we could from the historical, political, economic, and cultural perspective. At first, we thought it was because you had more powerful guns than we had. Then we thought it was because you had the best political system. Next we focused on your economic system. But in the past twenty years, we have realized that the heart of your culture is your religion: Christianity. That is why the West has been so powerful. The Christian moral foundation of social and cultural life was what made possible the emergence of capitalism and then the successful transition to democratic politics. We don't have any doubt about this.[49]

Joseph Needham, author of the monumental series *Science and Civilization in China,* may have influenced the Chinese academicians. Needham believed that the Chinese never developed modern science because they lacked the idea of unchangeable physical law, and they did not have this idea because they lacked the idea of an unchangeable Lawgiver, that is, a personal God.[50] Modern Chinese scholars have always been open to Needham's opinions, because Needham was a very rare Westerner: a Marxist, Maoist Christian.

In astronomy, the main intellectual barrier to the Copernican system was the belief, thanks to Aristotle, that the Moon, planets, and stars were subject to different laws than the Earth. Terrestrial objects were made up of different proportions of earth, water, air, and fire (the four terrestrial elements), while the heavenly bodies were made up of the fifth element, the quintessence. This fifth element, also called *ether,* was superior to the four mundane materials, because it underwent no change. A Christian philosopher, John Philoponus (490–570), challenged this view on the grounds that everything except God underwent change, and the heavenly bodies were no exception.[51] Everything in existence was subject to the same physical laws. This is essential to the Copernican system, because the Earth is the third planet from the Sun and hence must be subject to the same laws as the other planets. Unfortunately, by the sixth century, there were no astronomers alive with sufficient ability to resurrect Aristarchus's model and fit it to the Christian worldview.

However, Christianity has been considered an opponent of science because of the condemnation of Galileo for heresy by a Dominican tribunal. Indeed, Galileo was condemned, but the truth about what happened is almost the opposite of what is generally believed. Stillman Drake, the leading Galileo scholar of the past thirty years,[52] and Giorgio de Santillana, my own teacher and the leading Galileo scholar before Drake, have set the record straight. Drake has pointed out that Galileo's actions make no sense unless one first realizes that he was a Catholic zealot—the term used by Galileo and his friends; in current terminology, Galileo was a Catholic fundamentalist.[53] Like all fundamentalists, Galileo believed in Bible inerrancy, and he believed that the events described in the Bible make more sense in modern physics and Copernican astronomy than they do in the physics of Aristotle. The first chapter of Genesis, for example, has always been interpreted to say that the universe had a beginning, whereas in Aristotle the universe has always existed. This inconsistency made great difficulties for St. Thomas Aquinas, who wished to base Christian theology on Aristotle's physics. Galileo hoped to persuade the Church to adopt modern physics, but he was afraid the philosophers, whose jobs at the universities depended on the Church's acceptance of Aristotle, would try to prevent this change.

The philosophers (in modern terminology, scientists) succeeded. They arranged for Galileo to be tried for heresy, a crime of which he was innocent. We have the records of the trial.[54] The charge was made to Galileo, and he produced a document that would conclusively prove his innocence. The trial was immediately recessed. The next day Galileo confessed his guilt. Why?

The Galileo trial is similar to the war crimes trial of the Japanese general Hideki Tojo after World War II. The American plan after the war was to picture the Emperor Hirohito as the ignorant tool of powerful Japanese warlords. The Japanese people had been indoctrinated to believe they existed to serve the emperor, who was the symbol of Japan. By blaming the crimes committed by the Japanese armed forces on the warlords, the emperor could escape responsibility, and the Americans could rule Japan through the emperor. Fifty years later there is considerable evidence that Hirohito knew and approved of what his army was doing. But if this evidence had gotten out to the American people, they would have demanded that Hirohito be tried for war crimes also, ruining the American government's plan. In his trial, Tojo made the remark "Of course the Emperor knew what we generals were doing." A recess was immediately called. The next day Tojo testified that the emperor was kept in the dark about the army's war crimes. Almost certainly an American attorney told Tojo that if he testified the emperor knew and approved the war crimes, the emperor would be tried also and would probably be sentenced to hang. Tojo was going to be condemned whatever he said, but if he testified the emperor did not know, the emperor would be safe. Tojo was a loyal Japanese, taught to serve the emperor. By lying under oath, he could protect what he had sworn to protect, his emperor.

Galileo was a Catholic fundamentalist. If a Dominican privately told him that the Church's image would be harmed if he protested his innocence, he would confess to the charge of heresy, which in this case just meant that he had disobeyed an order not to discuss the Copernican theory. He confessed to this minor charge and suffered a nervous breakdown when he was sentenced to house arrest for life. In point of fact, the Catholic Church, and Christianity in general, would have been better off if he had proven his innocence. The importance of the Christian worldview for science would have been generally appreciated.

But if indeed the Christian worldview was responsible for the birth and growth of modern science, can modern science survive if Christian belief disappears? Atheism has in the past few decades replaced Christianity as the primary belief among the faculty at American research universities,[55] so we shall soon find out. In his 1937 inaugural address as incoming president of Yale University, Charles Seymour said, "I call on all members of the faculty, as members of a thinking body, freely to recognize the tremendous validity and power of Christ in our life-and-death struggle against the forces of selfish materialism. If we lose that struggle, judging from present events abroad, scholarship as well as religion will disappear."[56] In his 1951 book, *God and Man at Yale,* William F. Buckley, Jr., the founding editor of the conservative magazine *National Review,* claimed that Christianity was being deemphasized at Yale. The reaction of the Yale administration and faculty at the time was to deny it.[57] Fifty years later, is there any doubt that Christianity has disappeared as a significant force, not only at Yale but also at all the major American universities? Today it is hard to remember that Harvard was originally established to train Episcopal ministers; Princeton, Presbyterian ministers; Yale, Congregationalist ministers. The University of Chicago was once a Baptist university.

There are many disquieting signs that Yale's President Seymour was correct: with the disappearance of Christianity from the universities, scholarship is also disappearing. The decay of belief in an unchanging God is now being followed by the decay in belief in the existence of unchanging physical law underlying the natural world. The daily newspapers are full of absurd statements made by humanities and social science faculty at the elite universities, fully justifying George Orwell's statement that some ideas are so stupid only an intellectual could believe in them.[58] Irrationality in the humanities is often termed *postmodernism.* However, in this book I am concerned only with the effect of the decay of Christian belief on the natural scientists.

In 1962, the very year Richard Feynman discovered the correct theory of quantum gravity, the philosopher Thomas Kuhn (1922–1996) published his book *The Structure of Scientific Revolutions,* the first great attack on the idea that the physical laws exist. Kuhn's theory was and is enormously influential. I know: when I was a postdoc at Berkeley in the late 1970s, Kuhn filled the largest auditorium on campus when he came to lecture on his theory. He claimed that a scientific revolution occurs by the replacement of one "incommensurable" theory by another. The older theory is based on one "paradigm," or worldview, and the newer on another paradigm, and there is essentially no overlap between the paradigms of the two theories. Thus, the newer theory cannot be said in any sense to approach more closely the true laws of physics. The experimental evidence that supposedly persuaded physicists to replace the older theory with the new was actually incidental. The real reason physicists accepted the new theory was aesthetic—spiritual—namely, they found its paradigm more intellectually appealing. Since the newer theory does not approach reality more closely than the older theory, we cannot say that there is any evidence that true and fundamental laws of physics even exist.

Kuhn's examples were all taken from the (then) recent history of physics, primarily the replacement of classical Newtonian mechanics with quantum mechanics and general relativity. Forty years later, we can say with assurance that Kuhn did not understand the true relationship between the old and the new physics. I described that correct relationship in Chapter 2. Kuhn insisted on comparing classical mechanics in its simplified, more primitive single-universe formulation with the full multiverse formulation of quantum mechanics. This is comparing apples with oranges. To compare classical and quantum mechanics correctly, one must compare classical mechanics in *its* multiverse formulation—Hamilton-Jacobi theory—with quantum mechanics. Then and only then can one see that they are based on the identical paradigm, the multiverse. Similarly, to compare Newtonian gravity theory with Einsteinian gravity theory—namely general relativity—correctly, one must first formulate Newtonian gravity theory in its most powerful form: Cartan curvature theory. Then one sees that Newtonian gravity and Einsteinian gravity are based on the same paradigm: gravity is curvature.

Unfortunately, most physicists at the elite research universities are unwilling to accept the unique paradigm and the unique theory indicated by experiment and mathematical consistency. They want instead to impose their own aesthetic principles on physics, and to the Devil with experiments. Instead of the unique Theory of Everything discovered some thirty years ago, they insist that nature must obey "supersymmetry" theory, usually in the form of superstring physics, or brane physics. *Supersymmetry* is a mathematical transformation of bosons into fermions—of integer spin particles into half-integer spin particles—and vice versa. A necessary implication of a supersymmetric theory is that each particle we actually observe must have a "superpartner": for each boson—the gluon, say—there must exist a fermion with similar properties, the gluino. For each fermion—the electron, say—there must exist a boson, the selectron, with properties otherwise similar to those of the electron (expect for mass). The experimental problem with this proposal is that no supersymmetric particle has ever been detected. I have heard innumerable times, "We have discovered *half* of the particles predicted by supersymmetry, we only need to search for the other half." By *half* these "physicists" are referring to the known particles, not to their supersymmetric partners. By the same logic, we can say we have observed Cornish pixies. That is, we have observed their home, namely Cornwall. We need now only to search a little bit more before we observe the pixies themselves. In reality, there is no evidence whatsoever for pixies, and no evidence whatsoever for supersymmetry. If there is no supersymmetry, there are no supersymmetric strings, or branes. To argue that supersymmetry exists on the basis of mathematical beauty rather than on the basis of material experimental evidence is a secular version of the Gnostic heresy.

Of course, the superstring theorists deny that mathematical beauty is their main reason for working on supersymmetry. They claim an experimental justification, namely the absence of a consistent quantum gravity theory. This claim is nonsense. Richard Feynman discovered a consistent (*renormalizable* is the technical term) theory of quantum gravity forty years ago, and this theory is essentially unique. However, the superstring theorists find the Feynman theory "spiritually" unacceptable because it necessarily has a cosmological singularity. Thus we come to the real reason why many modern physicists find standard quantum gravity unacceptable: *it implies the existence of God!* If the existence of the Cosmological Singularity—God—is accepted, then it becomes mathematically possible to transform the renormalizable theory of quantum gravity into a theory that not only is term-by-term finite but, in addition, has a finite power series in the coupling constants. In effect, infinities that would otherwise occur in the laboratory are transferred to the Cosmological Singularity. In other words, God stabilizes the multiverse, thereby preventing it from collapsing into nonexistence. But for secularists, God must be eliminated at all costs. If necessary, they are willing to abandon experimental science itself.

But the most pernicious form of Gnostic dualism in modern science is not superstring theory but Darwinism. The idea of evolution in the sense of common descent is completely consistent with Christianity; theologians made this clear in the nineteenth century, and it has been confirmed in recent times by Popes John Paul II and Benedict XVI. A comparison of the DNA in chimpanzees and humans indicates that the two species had a common ancestor between 5 and 6 million years ago, and were we to see this common ancestor, we would probably call it an "ape." Furthermore, all metazoans—living beings made up of more than one cell—had a common ancestor, a single-cell organism, approximately 2 billion years ago. This fits nicely with the creation account in Genesis 2:7: "And God formed man out of the slime of the Earth." Being descended from an ape is better than being descended from slime, but indeed our one-cell ancestor was slime. However, Darwinism goes beyond the fact of common descent and claims that the mechanism was natural selection acting on "random" variation. There is no objection to natural selection, but the idea that evolution has no goal and is undirected—in short, is "random"—is an attack on Christian theology at its heart.

It is also an attack on the central foundation of physics. Newtonian physics was based on determinism: given the state of the universe at one time, the laws of physics would uniquely determine the state of the universe at all other times. However, as I pointed out in Chapter 2, the mathematical physicists of the late nineteenth century discovered that Newtonian physics in its most powerful form, Hamilton-Jacobi theory, was not fully deterministic as originally expressed. This difficulty was resolved in 1926 by Erwin Schrödinger, who added a term itself subject to a second equation to the Hamilton-Jacobi equation, and the pair of equations was equivalent to what we now call Schrödinger's equation, which is completely deterministic and, as a bonus, correctly describes the behavior of atoms and molecules. The history of physics, in other words, can be understood as the development of the full implications of determinism. As I emphasized in Chapter 2, determinism in quantum mechanics is called *unitarity,* and unitarity means that we can think of determinism as acting from the ultimate future backward in time. The evolution of matter is fundamentally teleological. What matter does in the present is constrained by the fact that it must evolve into the Omega Point, the Final Singularity, which is the First Hypostasis of the Cosmological Singularity.

In particular, unitarity requires that intelligent life must necessarily evolve independently on planets around stars several billion light-years apart in order that these intelligent life-forms can cancel the acceleration of the universe, which would otherwise destroy unitarity. The Cosmological Singularity in effect has always been directing the variations that have appeared in the genome of the biosphere, and has been directing which individuals actually mate. The term for this in Christian theology is *God's Providence.* Christians can never abandon trust in God's Providence. Physicists can never abandon trust in unitarity.

In the last two pages of his 1868 book, *The Variation of Animals and Plants Under Domestication,* Charles Darwin (1809–1882) eloquently described the central contradiction between his theory of evolution and the determinism of physical law, although he expressed the contradiction in terms of theology:

> And here we are led to face a great difficulty, in alluding to which I am aware that I am traveling beyond my proper province. An omniscient Creator must have foreseen every consequence which results from the laws imposed by Him. . . . If we assume that each particular variation was from the beginning of all time preordained, the plasticity of organization, which leads to any injurious deviations of structure, as well as that redundant power of reproduction which inevitably leads to a struggle for existence, and, as a consequence, to the natural selection or survival of the fittest, must appear to us superfluous laws of nature. On the other hand, an omnipotent and omniscient Creator ordains everything and foresees everything. Thus we are brought face to face with a difficulty as insoluble as is that of free will and determinism.[59]

We shall see in Chapter 11 how to resolve the conflict between free will and determinism, and Darwin showed great insight when he connected the resolution to the Problem of Evil. But Darwin is also correct in pointing out that his proposal for the mechanism of evolution, natural selection acting on "random" variation, is inconsistent with physical determinism of all events in the multiverse. I am in complete agreement with Albert Einstein, who said in response to the claim that there is a fundamental indeterminism, or "randomness," in nature: "That nonsense is not merely nonsense. It is objectionable nonsense."[60]

Darwinists are ultimately responsible for introducing this objectionable nonsense into physics by their unfortunately successful efforts to change the meaning of the word *probability.* The great French mathematical physicist Pierre-Simon de Laplace (1749–1827) made *probability* mathematically rigorous by defining it as a measure of human ignorance. In Chapter 2, I showed how probability arises in quantum mechanics, a consequence of human ignorance of the existence of the other universes of the multiverse. Probability was introduced in quantum mechanics by Max Born (1882–1970) in the 1920s. However, Born did not interpret probability as human ignorance. Instead, he believed it to be a frequency: if the spin, for example, of the electron were measured repeatedly, the probability that the spin is up is the ratio of the number of times it is measured to be up to the number of times it is measured. Furthermore, the true probability is measured only if the spin of the electron is measured an infinite number of times. It is possible, though improbable, that an electron whose wave function corresponds to being spin-up in half of the universes nevertheless measures spin-up in five consecutive measurements.

The problem with the frequency interpretation from the scientific point of view is obvious. It is not possible to carry out an infinite number of measurements. The human ignorance definition of probability does not have this difficulty. Pierre de Laplace, Carl F. Gauss, Augustin-Louis Cauchy, and Simeón-Denis Poisson, the four greatest mathematical physicists of the late-eighteenth and nineteenth centuries, developed the human-ignorance interpretation of probability. Why were Born and the other physicists of the early twentieth century unaware of their work?

Darwinism needed a different interpretation of probability. According to Darwin's theory of evolution, species evolve by natural selection acting on "random" variation. The words *random* and *chance* are synonyms for whatever it is that probability measures. Suppose that "probability" indeed was a measure of human ignorance. Then a typical Darwinian explanation would read: "Approximately 200 million years ago, mammals evolved from therapsid reptiles by means of natural selection acting on *human ignorance.*" Expressed in this way, Darwinism is obvious nonsense. But were "chance" seen to be an ultimate feature of reality, if such "chance" replaced "human ignorance" in the preceding sentence, then Darwinism would be a possible theory. So beginning in the middle of the nineteenth century, Darwinians went to work on the theory of probability. John Venn, Karl Pearson, and Sir Ronald Fisher created a new theory of probability, based on frequency and not human ignorance. Darwinism made sense in this new theory. Its creators, particularly Pearson and Fisher, were also convinced Darwinians. This frequency theory negatively influenced the development of physics because it delayed the acceptance of the multiverse interpretation of quantum mechanics for at least half a century.[61]

There is no "chance" in reality. The time evolution of the universe is unitary. This means that the multiverse has a goal, the universe has a goal, and each atom has a goal in the Ultimate Future. Since animals and plants are made up of atoms and are small parts of the universe, they also have goals. One goal of the biosphere on Earth is to give rise to intelligent life so it can expand out and cancel the acceleration. In Chapter 11 we shall see the purpose of the other life-forms, those that have no obvious role in generating intelligent life. Cardinal Christoph Schönborn, the archbishop of Vienna, in a controversial 2005 *New York Times* op-ed piece, defended the traditional Christian view that evolution of body forms is consistent with the Bible—there is nothing unchristian about chimps and humans having a common ancestor 5 million years ago—but Darwinism is not: "An unguided evolutionary process—one that falls outside the bounds of Divine Providence—simply cannot exist." Physics says exactly the same: there is nothing in reality that is outside unitary time evolution: Final Cause. To accept chance as an ultimate is to accept human ignorance as an ultimate. In the words of Cardinal Schönborn, "In the modern era, the Catholic Church is in the odd position of standing in firm defense of reason as well. . . . Scientific theories that try to explain away the appearance of design as the result of 'chance and necessity' are not scientific at all, but as John Paul put, an abdication of human intelligence."[62]

Can science itself survive an "abdication of human intelligence"? If science was born from a Christian worldview, can it survive if a Christian worldview disappears from people who hold the title "scientist"? There is considerable evidence that it cannot. We have seen that Greek science disappeared in 100 B.C. even though Greeks with the title "scientist" (natural philosopher or mathematician) continued for several hundred more years. Superstring theory has replaced experimental physics in physics departments worldwide. The life expectancy of a white male who reaches the age of seventy in the Western countries has increased only two years since 1950, in spite of a huge increase in expenditure on medical research.[63] If "scientists" no longer believe in an unchanging natural order, created by an unchanging God, they will no longer search for laws that they no longer believe exist.

A final manifestation of the Gnostic heresy I wish to refute is the idea that religion and science belong in distinct categories: religion is concerned with moral questions and science with factual questions. Religion is concerned with the spiritual world and science with the material world, in other words. Obviously, this is the ancient Gnostic dualism reborn. Like the ancient Gnostics, many moderns who espouse this view argue that it is blasphemous to use science to prove the existence of God, or to justify any particular religion—Christianity, say. These claims are nonsense on several levels.

People who assert that religion and science are to be kept strictly separated generally also say that the two are separate but equal. I grew up in Alabama in the 1950s, when blacks and whites were racially separated by law. "Separate but equal" was the slogan used by the segregationists of my home state at the time. But I knew from my own experience that the actual reality was separate but very unequal. Similarly, those who advocate "separate but equal" in segregating science and religion really mean that religion should be kept out of science because religion is factually false. These people truly believe that God does not exist and hence does not have any effect on reality. Those who claim that religion is concerned with morality are the same people who tell Christian leaders, especially the Roman Catholic bishops, to stick to religion when the Christians express their moral opposition to abortion. In other words, Christianity has nothing to say about anything that happens in material reality. This would be true if the Gnostics were correct that there is a spiritual reality outside the control of the god who created the material universe. But it is not true, since the God Who created the material world also created the spiritual world, and the latter is based on the former, as discussed in Chapter 3. I shall show in Chapter 7 that moral questions are ultimately factual questions: if one knew all the facts—only God Himself can know all the facts—then there would be no argument over moral questions.

The claim that it is blasphemous to attempt to establish religious truth by science, particularly by scientific experiment, is refuted by a glance through either the Old or the New Testament. In 1 Kings, the prophet Elijah proposed to the people of Israel that they put to the experimental test which god, Baal or Yahweh, is the true God: "Let them therefore give us two bullocks; and let them choose one bullock for themselves, and cut it in pieces, and lay it on wood, and put no fire under: and I will dress the other bullock, and lay it on wood, and put no fire under. And call ye on the name of your gods, and I will call on the name of Yahweh, and the God that answers by fire, let him be God. And all the people answered and said, It is well spoken" (18:23–24).

The people of Israel 3,000 years ago had more sense than many scholars today. If Yahweh cannot "answer by fire"—perform miracles today—then He is not God. If His existence and actions in the worlds cannot be seen by science, then He does not exist. The Gospels emphasize that Jesus was considered to be from God because of the miracles He performed, and the same was true of His apostles: "Then Philip went down to the city of Samaria, and preached Christ unto them. And the people with one accord gave heed unto those things Philip spoke, hearing and seeing the miracles he did. For unclean spirits, crying with loud voices, came out of many that were possessed with them; and many that were taken with palsies, and that were lame, were healed" (Acts 8:5–7). These miracles are the same miracles that are now converting many throughout the world. People now, as 2,000 years ago, are being converted by seeing with their own eyes the action of God in the material world. To say these acts are blasphemous is to say Christianity and Judaism are blasphemous.

A miracle must be more than an improbable event interpreted by us as an act of God. It must in addition be *proven* to be an act of God. In the most famous debate of the eighteenth century, between the two greatest scientists of the time, Sir Isaac Newton (1642–1727) and Gottfried Leibniz (1646–1716), this point was emphasized by Newton:

> The Notion of the World's being a great *Machine,* going on *without the interposition of God,* as a Clock continues to go without the Assistance of a Clockmaker; is the Notion of *Materialism* and *Fate,* and tends (under pretense of making God a *Supra-Mundane Intelligence*) to exclude *Providence* and *God's Government* in reality out of the World. And by the same Reason that a *Philosopher* can represent all Things going on from the beginning of the Creation, *without* any Government or Interposition of Providence; a *Sceptick* will easily Argue still farther Backwards, and suppose that Things have from Eternity gone on (as they now do) *without* any true Creation or Original Author at all, but only what such Arguers call *All-Wise and Eternal Nature.* If a *King* had a *Kingdom,* wherein all Things would continually go on *without* his Government or Interposition, or *without* his Attending to and Ordering what is done therein; it would be to *him,* merely a *Nominal* Kingdom; nor would he in reality deserve at all the Title of King or Governor. And as those Men, who pretend that in an Earthly Gov-

> ernment Things may go on perfectly well *without* the *King himself* ordering or disposing of any Thing, may reasonably be suspected that they would like very well to set the King aside: So whosoever contends, that the Course of the World can go on *without* the Continual direction of *God,* the Supreme Governor; his Doctrine does in Effect tend to Exclude God out of the World.[64]

This passage is taken from the opening of what has become known as the "Leibniz-Clarke correspondence," because although it was really a debate between Newton and Leibniz, the former did not appear in his own name. The English theologian and philosopher Dr. Samuel Clarke instead represented Newton. The debate, in the form of an exchange of letters between Clarke and Leibniz, began when the latter sent a letter to Caroline, Princess of Wales, complaining that the study of natural theology had declined in England as the result of the pernicious influence of Sir Isaac Newton's physics. Newton could not afford to ignore this attack, because it threatened not only his work as a scientist but also his livelihood as a government official. Having earlier charged Leibniz with plagiarizing his invention of differential and integral calculus, Newton refused to debate Leibniz directly; his friend Clarke stood in his place. But Princess Caroline, who knew all three men, assured Leibniz, "You are right about the author of the [letters of] reply; they are not written without the advice of Chevalier Newton."[65]

Newton believed that his law of universal gravitation showed the solar system to be unstable, and by his calculations, the solar system would fly apart about 10,000 years after being set in motion. Since this number was approximately equal to the time since the creation of the world accepted then, 4000 B.C., Newton thought his physics had verified the traditional date for the creation of the universe. The French mathematical physicist Pierre Laplace showed in the late 1700s that Newton had ignored some crucial terms in his calculation, and when a few additional terms were taken into account, the solar system could be shown to be stable. Unfortunately for Laplace, the English mathematical physicist John Couch Adams showed in the late 1800s that Laplace himself had ignored a few terms himself, and if even more terms were considered, the question of the solar system's stability was once again put in doubt. Even today, with our supercomputers to help us with the calculations, we don't know if the solar system is stable under Newtonian gravity. But we do know that if the solar system is unstable, it is nevertheless stable for far longer than the 10,000 years that Newton defended.

There are several lessons in this story of stability calculations. The first is that Newton truly believed he had established using standard physics that God had directly intervened in the world at the traditional time given for the creation of the universe, approximately 4000 B.C. Let me emphasize that this traditional date is not restricted to Christians. The Jewish calendar's first year is computed to be the first year the universe existed, and this year is held to be 3760 B.C., so the year 2007 is the year 5767 in the Hebrew calendar.[66] The second lesson is that Newton believed this instability established a breakdown in the known laws of physics. But this was an error, not only a mathematical error but also an error from the traditional Christian view of miracles: God *never* sets aside fundamental laws of nature. The laws of God are never violated, only our human understanding of what these laws actually are.

Miracles of Conversion

Christianity is the religion founded on a miracle—the Incarnation—and justified by reference to a miracle—the Resurrection. Even today, conversions to Christianity occur mainly through miracles, the same miracles that are described in the New Testament: raising the dead, healing the sick, casting out demons, and seeing visions of Jesus or angels. In the United States, most conversions occur in the form of "born again" experiences, which are versions of the vision experienced by St. Paul on the road to Damascus. According to the leaders of the "house churches" of China, "As many as 80 percent of believers first come to Jesus because they receive a miraculous healing or deliverance from the Lord."[67] In house churches, the parishioners meet in someone's house rather than in a formal church building. This is the main form of Christianity in nations where Christian worship is illegal or strictly controlled by the government, as it is in China today. In colonial Virginia a few decades before the American Revolution, many people—for example, Patrick Henry—attended house churches because all denominations other than the official Church of England were banned. To get a feel for how important miracles of healing are in the growth of Christianity, I advise you to check out books on the experiences of Chinese apostles—for example,

Back to Jerusalem or *The Heavenly Man: The Remarkable True Story of Chinese Christian Brother Yun,* both by Paul Hattaway. I call the people described in this book "the Chinese apostles" because their stories sound almost exactly like those recorded in the New Testament's Acts of the Apostles.

Let us now consider how the main miracles of conversion today might work in a manner consistent with the laws of physics. There are four main miracles of conversion:

1. Raising the dead
2. Healing the sick
3. Casting out demons
4. Seeing visions

Raising the Dead. First, we need a definition of the word *dead.* The image of a dead man in most people's minds is a human skeleton, but it is exceedingly rare that Christian apostles put flesh on the bones of skeletons. God could put flesh on a skeleton without violating the laws of physics, by the mechanism that I describe in Chapter 8, but He rarely does so. All accounts I have been able to find indicate that the people who are raised from the dead have been dead no more than three days.[68] The body is identified as "dead" by ordinary people with no medical training. Even doctors can make mistakes when declaring a person to be dead. A person's heart can stop for a short period and yet be restarted, by an electrical shock or even spontaneously. The fact that death can be misdiagnosed and a person buried only later to reanimate in a coffin is what led nineteenth-century legislators to pass laws requiring the "dead" to be embalmed before burial. Embalming entails replacing the blood with embalming fluid. This procedure makes certain that the body in the coffin is indeed dead. Before embalming became standard procedure in the United States, some exhumed coffins had scratch marks on the tops from the fingernails of "corpses" that woke up inside the buried coffins.

If a "dead" body suddenly wakes up after being prayed over in full view of an assembled body of skeptics who are convinced that the person is dead, the skeptics often cease to be skeptics. If indeed prayer had the effect of waking up the "dead" body, then the Christian apostles can take the credit. No experiment has been carried out to determine if prayer can raise the dead. One could, for example, compare the number of reports of the "dead" spontaneously waking without prayer with the number of reports in which this occurs with prayer. However, there is no report of the dead being raised that is not consistent with a person in deep suspended animation suddenly being reanimated. No physical law is violated in such a case.

Healing the Sick. Examples of healing the sick include the healing of almost any disease one can name. However, the healing occurs after the doctor has given the patient up for lost or when the sick person is too poor to afford medical treatment. The patient, after being prayed over, is observed to recover spontaneously. The cancer tumor disappears, the bacteria of the infection disappear, the lame walk. If a person is partially or totally convinced the prayer will work, there is a chance the person's conviction will be communicated to his or her body. In conventional medical parlance, this is called the *placebo effect.* In any test of a new drug, researchers know that giving patients a pill containing nothing but sugar will cure a statistically significant fraction of the sick. The sugar pill does nothing; the patient's mind does everything.

There is at present no known way of separating placebo-effect cures from prayer-effect cures. However, there is no example of a cure occurring after a prayer that is inconsistent with physical law. One of the healing miracles attributed to the saint at Lourdes was a broken bone that healed "instantaneously." But the healing times of a broken bone follow a Gaussian (bell-shaped) distribution, with a mean time of several weeks. The distribution means that the healing time will vary from individual to individual. It is not physically impossible for a bone to heal in a day or less, just gigantically improbable by human ways of measuring probability. Once again, a miracle is an event allowed by natural law but improbable according to human knowledge. Recall that "probability" is a measure of human ignorance, not human knowledge.

Casting Out Demons. When I am asked, "Do you really believe in demons?" I reply, "Do I believe in computer viruses? Yes, I most certainly do." In other words, demons exist, but they should be thought of as forms of computer viruses running on the computer that is the human brain. It scarcely needs emphasizing that Christians have no choice; we must believe in demons. Jesus, according to all the Gospels, spent much of his time casting them out of people who had been possessed. If demons do not and did not exist, what was He doing? Computer viruses are small programs, generally sent from one computer to another via e-mail, that take control of the computers in which they find themselves. A personal computer infected by a virus still employs its basic operating system (most often some version of Microsoft Windows), but the memory and processing hardware of the computer are no longer used for the purposes of the computer's owner. Rather these resources are used to carry out the purposes of the virus (in present-day computer viruses, these purposes are usually making copies of the virus in the memory of the personal computer and sending out additional copies to all e-mail addresses the virus can find in the computer's memory). In a few years, computer experts expect computer viruses to become more sophisticated and take over the infected computer and use it to perform more complicated computations.

In human brains, the equivalents of computer viruses are called *multiple personality disorder* (MPD) or, in recent times, *dissociative identity disorder.*[69] People with this mental disease appear to have several personalities in the same body. In the last century in the West, the different personalities have considered themselves merely different human personalities, but in the past, many of these personalities claimed to be supernatural beings: demons.[70] There are several "cures" for MPD, all of which involve suppressing all but one of the personalities. The "central" personality—usually the personality the clinical psychologist treating the patient finds most congenial—is persuaded to ignore the other personalities. If the alternate personalities are never allowed to "run" on the human brain, they eventually disappear. Effectively, the computer virus that corresponds to the alter ego is deleted. All reports by Christians who claim to have cast demons out of "possessed people" that I have been able to find are consistent with the demons being manifestations of MPD.[71] If one or more of the personalities claim to be a demon and the central personality is persuaded (consciously or unconsciously) that the religious ceremony will destroy the demonic personality, then it probably will. As in raising the dead and healing the sick, no physical laws need be broken in order for demons—the computer viruses of MPD—to exist and to be cast out.

Christianity, of course, claims that there is a chief demon—Satan—who is the master of the lesser demons. In Chapter 7, I shall suggest that such an entity does indeed exist, but that he is a computer virus not in our brains but in our DNA, and indeed in the DNA of most metazoans. He can manifest himself in our brains only by generating evil behaviors, one of which is the possibility of MPD. In such a case, Satan not only exists but indeed is the master of the lesser demons.

Seeing Visions. A classic example of a vision miracle is the Annunciation: the angel Gabriel appeared to Mary to announce that she would bear a son though she had never known a man.

In South Africa, 42 percent of new converts from Islam accept Christianity because they see visions in which they are told that Jesus is God.[72] The son of the famous atheist Madalyn Murray O'Hair became a Baptist minister after seeing a vision of an angel.[73] In all cases, those who see visions of angels are first exposed to the Christian belief that such visions are to be expected. No skeptic would have difficulty accepting the possibility of autosuggestion—the belief that one *can* see a vision itself generates the vision—but, of course, no convert to Christianity interprets the vision this way. The cumulative evidence for Christianity in

Figure 5.1. *The Annunciation* by Henry Ossawa Tanner. The Annunciation, Mary being informed by the angel Gabriel that, though a virgin, she would bear a son, is an example of a "vision" miracle. Like all miracles, vision miracles do not in any way violate physical laws.

the following chapters will show that they are probably correct, and the skeptics wrong. The God who has resurrected His Son to show us how to use that power needs Christians in plenty for this purpose.

Christians claim that the dead are raised, the sick are healed, the demons are cast out, and angels appear not by any power they possess but by the aid of Jesus, in Whose Name these miracles are performed. I shall show in Chapter 9 how this could in fact be true, and how to test experimentally if it is.

In the following chapters I shall show that the Virgin Birth, the Resurrection, the Incarnation, and all of Jesus' "nature" miracles were miracles in the orthodox Christian sense. All are manifestations of the direct action of God in the material world, not in violation of physical law but in conformity to it. Not against God's Word but in conformity to God's Word: physical law. I shall show exactly how these miracles are completely consistent with known physical law.

VI

The Christmas Miracle: The Star of Bethlehem

Now after Jesus was born in Bethlehem of Judea in the days of Herod the king, Magi from the east arrived in Jerusalem, saying, "Where is He who has been born King of the Jews? For we saw His star in the east and have come to worship Him." When Herod the king heard *this,* he was troubled, and all Jerusalem with him. Gathering together all the chief priests and scribes of the people, he inquired of them where the Messiah was to be born. They said to him, "In Bethlehem of Judea; for this is what has been written by the prophet: 'And you, Bethlehem, land of Judah, are by no means least among the leaders of Judah; for out of you shall come forth a ruler who will shepherd my people Israel.' " Then Herod secretly called the Magi and determined from them the exact time the star appeared. And he sent them to Bethlehem and said, "Go and search carefully for the Child; and when you have found *Him,* report to me, so that I too may come and worship Him." After hearing the king, they went their way; and the star, which they had seen in the east, went on before them until it came and stood over *the place* where the Child was. When they saw the star, they rejoiced exceedingly with great joy.

MATTHEW 2:1–10 (NEW AMERICAN STANDARD BIBLE)

Then when Herod saw that he had been tricked by the Magi, he became very enraged, and sent and slew all the male children who were in Bethlehem and all its vicinity, from two years old

> and under, according to the time which he had determined from the Magi.
>
> MATTHEW 2:16 (NEW AMERICAN STANDARD BIBLE)

I SHALL ARGUE THAT IF THIS ACCOUNT IN MATTHEW'S GOSPEL is taken literally, then the Star of Bethlehem must have been a Type Ia supernova or a Type Ic hypernova, located either in the Andromeda Galaxy or, if Type Ia, in a globular cluster of this galaxy, pictured in Figure 6.1.

The account in Matthew may give the location of the supernova remnant to within an arc minute, at least in declination. Let me explain what these perhaps unfamiliar astronomical terms mean. An *arc minute* is a measure of size in the sky. The full Moon is about 30 arc minutes in angular size. Since a star is a point of light on the sphere of the heavens, two numbers must be given to describe the star's location on the heavenly sphere, just as two numbers must be given to describe the location of a point on the Earth's surface. On the Earth's surface, the two numbers are latitude and longitude. Astronomers take the latitude and longitude grid on the Earth and project it into the sky. *Declination* is what astronomers call the celestial latitude. *Right ascension* is the term astronomers give to the celestial longitude.

If indeed the Star of Bethlehem were a supernova in the Andromeda Galaxy, it would leave an expanding gas cloud as a remnant of the supernova explosion that occurred to announce the birth of Jesus. From the known rate that gases expand outward from a supernova explosion, we can calculate that this remnant would be 6 arc minutes in radius today and could be searched for in the Fe I resonance absorption line at 3,860 angstroms. (*Fe I* just means a normal, iron atom that has all its electrons. *Absorption line* just means a dark line appearing on the spectrum.) An emission line with this wavelength would appear to have the color violet if it were intense enough to be seen, but after 2,000 years, all energy emission has faded to invisibility to human eyes. Supernova 1885 (S Andromedae) was first observed in 1989 as an absorption-line extended object at this wavelength.[1] This observation as an extended object was more than 100 years after this supernova was first observed, as one can tell by the name. Astronomers label supernovas by the years they are first observed. Thus, S Andromedae was first seen as an explosion in the year 1885.

There have been many, many proposals made for what the Star of Bethlehem actually was. Some people have proposed that it was not an actual star but a comet. This is superficially plausible, because comets were believed by many people in the ancient world to announce the births of kings. But a comet is not a star. Other people have argued that the "Star" was a *conjunction*—this word means "very close approach on the celestial sphere"—of two or more planets. Once again, a conjunc-

Figure 6.1. Andromeda Galaxy. The Star of Bethlehem was a supernova or a hypernova in this galaxy. At peak luminosity, the supernova was as bright as the rest of the entire galaxy.

tion of two planets is not a star. Let me be blunt. In this chapter, I am going to concentrate on two aspects of Matthew's description of the Star of Bethlehem that I consider of vital importance. First, I am going to assume the word *star* means exactly that. The Star of Bethlehem is a star. It is not a planet, or a comet, or a conjunction between two or more planets, or an occultation of Jupiter by the Moon. I shall assume that the Star of Bethlehem was an actual point of light *fixed* on the celestial sphere. Second, I am going to assume that Matthew's expression "stood over" means exactly that. The star went through the zenith at Bethlehem. It did not hover vaguely in the general southerly direction as seen from Jerusalem. The expression "stood over" is not some wishy-washy term applied to any old comet that happened to be close to the horizon around A.D. 1. It is specific. It applies to the Star of Bethlehem, and the Star was there, in the sky, directly above the Magi, at the time of their visit to the baby Jesus.

More precisely, Matthew 2:2, quoted at the start of this chapter, refers unequivocally to a "star," and as the biblical scholar Raymond E. Brown has pointed out, this can mean only a nova, supernova, or a hypernova if taken literally.[2] The astronomer D. W. Hughes has emphasized that Matthew 2:9—"the star which they [the Magi] saw in the east [in the first rays of dawn] went before them, till it stood over where the young child was"—if taken literally, means that the Star of Bethlehem must have passed through the zenith at Bethlehem.[3] Since the latitude of Bethlehem is 31°43' north, the declination of the Star in the first decade B.C. (the range of estimates of Jesus' birth year) must have been 31°43' north.

Matthew 2:9 also suggests the right ascension of the Bethlehem Star. Matthew 2:2 tells us that the Magi were astrologers who believed in a correspondence between celestial and terrestrial events. It is plausible that they would identify the zero of right ascension—the vernal equinox—with what they would regard as the natural zero of longitude. The zero of longitude is usually taken as the longitude of the main observatory (we use the location of the Greenwich observatory, though in the nineteenth century the French unsuccessfully fought to use the location of the Paris observatory).[4] The central observatory of the ancient world was in Babylon, and indeed a Persian map has Babylon as the zero of longitude.[5] Bethlehem and Babylon have longitudes of 35°12' east and 44°26' east, respectively. Setting Babylon as the zero of longitude and identifying it with the zero of right ascension would give the right ascension of the Star of Bethlehem as 23 hours, 23 minutes, in 5 B.C.

This position in the first decade B.C. is far from the galactic plane (the likely location of a galactic nova or supernova), but it is very close to the Andromeda Galaxy, whose center in 5 B.C. was $(\delta, \alpha) = 30°13', 23^h, 1^m)$, where δ is the declination and α is the right ascension. The galactic halo of the Andromeda Galaxy would definitely have included the declination of the zenith of Bethlehem. The right ascension of the Andromeda Galaxy center would correspond to a position in the Mediterranean Sea, but the nearest large city with the indicated declination-latitude is Jerusalem, the city to which the Magi first traveled. The nearest small city is Jaffa, the main port of Palestine and, in Greek mythology, the home city of Andromeda, princess of Jaffa. Any astronomer of the first decade B.C. would immediately have associated an event in the constellation Andromeda with Palestine. Our system of constellations is essentially that of Ptolemy, which can be traced back at least to Eudoxus of Cnidus (c. 350 B.C.) through the poet Aratus, before the Seleucid period of Greek rule over Babylon.[6] Astronomical techniques at the time were sufficiently accurate to allow observers to determine that a star's declination was at the zenith of a given location to within a minute of arc, or within a nautical mile, using a dioptra and a plumb bob.[7] A supernova in M31 (the Andromeda Galaxy) could indeed have "stood over" Bethlehem.

Matthew's statement that the Magi first observed the Star "in the east" can also be translated "at helical rising" (in the first rays of dawn), as pointed out by Hughes.[8] I suggest that both interpretations are appropriate. Andromeda appeared in the eastern sky only in the late winter and early spring months in the first decade B.C. We would expect naked eye observers to have noticed a fifth-magnitude star (see later) only if they happened to be concentrating on that part of the sky, so a date near the vernal equinox is suggested. Also, such a faint star would likely have been seen only if it was in the east well up in the sky at dawn. On March 22, 8 B.C., the day after the equinox, there was a conjunction of Mars with the Sun. On this date, Venus was in the constellation Aries and located at its rising almost exactly below M31 and only about 20 degrees in azimuth north of east at Babylon. Venus on that day rose shortly after dawn and had magnitude −4.2, so it would have been visible after sunrise. On the local Babylon horizon, an observer would have used the stars in Andromeda as guides for the soon-to-rise Venus and would have seen the supernova in Andromeda, in the east, in the first light of dawn.

Ptolemy, who is best known today for his *Almagest,* the standard textbook on geocentric astronomy, the theory of the solar system that was replaced in the seventeenth century by Copernicus's heliocentric astronomy, was better known in the ancient world for his astrological treatise *Tetrabiblos.*[9] In this standard textbook of ancient astrology, Ptolemy associated every nation with a zodiacal constellation and a planet, and he associated Judea with Aries and Mars. He also associated Andromeda with a single planet, Venus.[10] Thus, on this date Andromeda would have been triply associated with Judea. Ptolemy associated the births of kings with the presence of Saturn, Jupiter, and Mars in a bicorporal sign (Gemini, Virgo, Sagittarius, and Pisces).[11] On March 22, 8 B.C., Mars, the Sun, and Saturn were in Pisces. Jupiter was in Aquarius, but Ptolemy assigned Jupiter to both Sagittarius and Pisces,[12] and he asserted that both the Sun and Jupiter rule the northwest triangle. So the Sun in Pisces can act as a stand-in for Jupiter. In 8 B.C., Augustus ordered a census of all Roman citizens,[13] so this year agrees with Luke 2:1. The great nineteenth-century archaeologist Sir William M. Ramsay provided additional arguments for 8 B.C. as Jesus' birth year.[14] Hughes also associates this 8 B.C. census with the census mentioned in Luke.[15] But none of these scholars points out the truly fascinating implication of this association: it would mean that Joseph, and hence Jesus as Joseph's son, was a Roman citizen. Thus, Jesus could have avoided the scourging and crucifixion, since both punishments were forbidden to be used on Roman citizens. Jesus went willingly to his horrible death, as claimed in traditional Christian theology.

Ptolemy asserted that the vernal equinox is present at the beginning of nativities, so a supernova at this equinox would naturally suggest a birth.[16] Hughes points out that June 20 is one of three traditional dates for Jesus' conception.[17] Nine months later is March 20. Hughes also emphasizes that March and April are the lambing season, and the most common time for shepherds to be with their flocks.[18] Thus, March 22 would be consistent with Luke 2:8. A date slightly later than the vernal equinox—one of the two solar points associated with a "birth"—would render understandable the Church's subsequent decision to move Jesus' birthday to a date slightly later than the other solar date associated with a birth: the winter solstice.

The March 22, 8 B.C., date for Jesus' birth is consistent with the date of his crucifixion, which can also be fixed by astronomy. Bradley Schaefer points out that the Jewish lunar calendar fixes the Passover date, and this in turn requires the crucifixion to have occurred on either April 7, A.D. 30, or April 2, A.D. 33.[19] However, in Acts 2:20, Peter quotes Joel 2:31 word for word: "The sun shall be turned into darkness, and the moon into blood, *before* that great and notable day of the Lord come" (my emphasis). Astronomers J. C. Humphreys and W. G. Waddington remark that "moon into blood" is often used to describe a lunar eclipse, and a lunar eclipse visible from Jerusalem occurred on December 9, A.D. 29.[20] The astronomer Fred Espenak has calculated that a total solar eclipse occurred just two weeks earlier, on November 24, A.D. 29, with Palestine in the penumbra and within 2 degrees of the umbra, as illustrated in Figure 6.2.[21] Modern calculations of the exact location of the umbra (where the solar eclipse appears total) are only accurate to one degree, so the eclipse could have been total to people living in Galilee. Everyone in Palestine would have seen at least a partial eclipse that was very close to a total eclipse.

Matthew 27:45 (cf. Luke 23:44; Mark 15:33) records a "darkness over the land," and a near contemporary, Thallus, writing about A.D. 52, asserts this "darkness" recorded in Matthew, Mark, and Luke was in fact a solar eclipse.[22] Luke 23:45 says explicitly that the Sun was darkened. A solar eclipse followed two weeks later by a lunar eclipse is very rare and would explain the impact of Peter's words on his listeners a few months after the crucifixion on April 7, A.D. 30: they would be impressed that the claim of Jesus' Resurrection—the day of the Lord—came just *after* the two eclipses that everyone in Palestine saw. Joel 2:10 asserts that an earthquake would occur before the sun and moon darkening, and a contemporary Greek, Phlegon, records that an earthquake, felt all over the Near East, also occurred in A.D. 29.[23] Matthew 27:51 and Luke 23:45 record such an earthquake. (The corresponding verse in Mark, 15:38, does not mention an earthquake, just the rending of the Temple Veil.) We are told in Luke 3:23 that Jesus began his public ministry when he was "about thirty years of age," which would mean that he was between twenty-five and thirty-five when he was crucified. His public ministry lasted between one and three years. A birth date of March 22, 8 B.C., would mean he was thirty-four years old in A.D. 27, consistent with a three-year ministry, ending in A.D. 30.

The brightest novae have an absolute visual magnitude of −9, so a nova in the Andromeda Galaxy would be invisible to the naked eye. A Type Ia supernova and a Type Ic hypernova have maximum absolute visual magnitudes of −19.5 and −19.4, respectively,[24] which would cor-

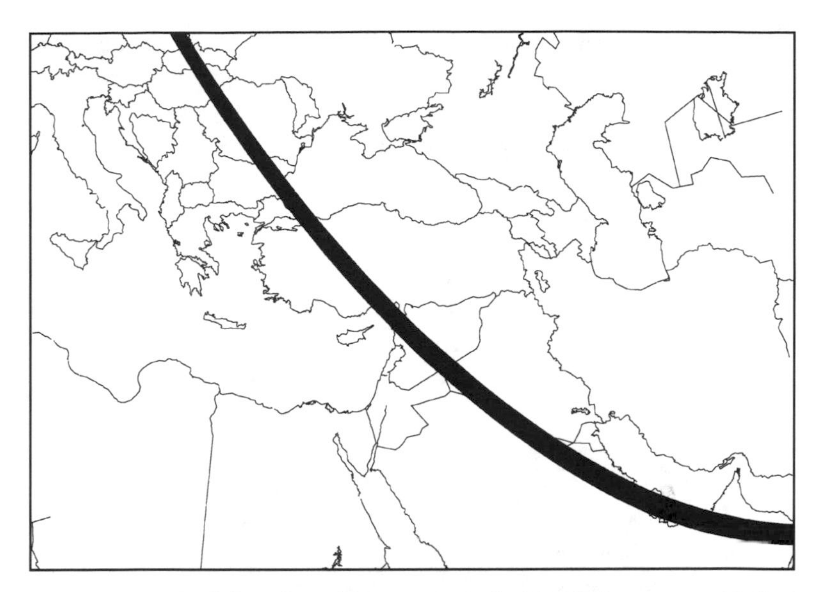

Figure 6.2. Path of the solar eclipse on November 24 in the year A.D. 29, four months before Jesus' crucifixion. The narrow black band is the path where the Sun is totally covered by the Moon. Within this band, people would see a total eclipse. People elsewhere on the map would see a partial eclipse. The inhabitants of Jerusalem would see an eclipse that was very close to total. Since the accuracy of modern calculations of the locations of ancient eclipses is accurate only to within one degree, it is possible that people in Galilee saw a total eclipse.

respond to an apparent visual magnitude of 4.5 at a distance of 750 kiloparsecs, the distance to the center of the Andromeda Galaxy. The subluminous supernova 1885 (usually called S Andromedae), the first Type I supernova observed in M31 (the usual astronomical name for the Andromeda Galaxy), had a peak visual magnitude of 5.8.[25] Supernova 1987A was initially identified by Ian Skelton with the naked eye when it had an apparent visual magnitude of 5.[26] The limit of naked-eye observation is an apparent visual magnitude of 6. But such a faint "new star" would be noticed by very few (most ancient recorded supernovae have an apparent magnitude less than −2), consistent, as pointed out by D. W. Hughes, with the statement in Matthew that Herod and his court were unaware of the "new star."[27] Hughes also mentions a tradition that the "new star" disappeared and was only rediscovered by the Magi observing the Star from a well (or cave).[28] A supernova will rapidly dim, and if the Magi took two weeks to reach Bethlehem—which is physically possible, as Ormand Edwards has pointed out[29]—this could be easily explained. There is a tradition that the Magi arrived in Bethlehem twelve days after first sighting the Star. The tradition that the Magi rediscovered the Star by seeing it in a well in Bethlehem[30] could also explain how they could determine that the Star passed through the zenith at Bethlehem even though it passed through the zenith in the daytime, which it would do if indeed it rose just before dawn, as I indicated earlier.

But there is another possible astronomical meaning of the Greek phrase *en te anatole,* translated into English as "in the east." As pointed out by Hughes in his book *The Star of Bethlehem,* the Greek *en te anatole* can also have the translation "acronychal rising," which means rising in the east just as the Sun sets.[31] If this translation is used, then another date for the Star is indicated: the autumnal equinox in 6 B.C. Hughes, M. R. Molnar, and M. Kidger have emphasized that astrologers would naturally associate the triple conjunction of Jupiter and Saturn in Pisces in 6 B.C. with the birth of a King of the Jews.[32] Pisces is associated with the Jews, and Jupiter is associated with kingship in Babylonian astrology. In Jewish astrology, Saturn was regarded as a protector of Israel. For the Babylonian astronomers, the two equinoxes were the most important dates of the year, so they would have been especially focused on the astronomical events occurring on both the eastern and western horizons on these two dates. The second of the triple conjunctions is very close to the autumnal equinox (September 29 in 6 B.C., according to Hughes).[33] Jupiter and Saturn rose about 5:45 P.M. in the Babylon sky on September 21, 6 B.C.

Hughes, Molnar, and Kidger have argued that this triple conjunction is the "Star" of Bethlehem. But as I said earlier, a conjunction, or near approach of two planets, is not a star. Stars are single points of light, not groups. Also, look again at the wording Matthew uses in the meeting between Herod and the Magi. The words strongly suggest that the Magi had just seen the Star, and that the Star was unexpected. Herod had neither seen nor expected the Star. However, Babylonian astronomers had developed the ability to predict the dates of planetary conjunctions a few years in advance. The triple conjunction would not have been unexpected. A supernova is unexpected even for astronomers today.

But Hughes and company have made an important point in calling attention to the triple conjunction and to the astrological associations of its location, which would focus the attention of the Magi on Judea and suggest the idea of the birth of a king in that land. Since there were several conjunctions of Jupiter and Saturn in the two years preceding the date I propose for the supernova's first appearance, the fact that conjunctions *could* suggest the birth of the King of the Jews provides an eminently reasonable explanation of why Herod had all the male children under the age of two killed (Matthew 2:16). He didn't want to take a chance that the supernova gave the exact date of the birth. As far as Herod could tell, the supernova might have occurred just to call the attention of the Magi to the conjunction. Hughes and company believe the conjunction was sufficient to announce the birth. There is a good chance that Herod, 2,000 years earlier, realized this indeed was a possibility.

There is a tradition in China that one of the Magi was Liu Shang, chief astrologer to the Chinese emperor at the time of Jesus' birth.[54] It is claimed that Liu disappeared for two years after discovering a "king star," an astronomical announcement of the birth of a king. This is possible if Liu computed the conjunction of Jupiter and Saturn, and somehow associated it with the birth of a king in the far west of China. It was possible to travel the Silk Road from China to Babylon, the "western" center of astronomy at that time. It would have taken a year to make this journey, and Liu could have arrived roughly just before the conjunction. He would, in other words, have arrived just as the light from the supernova reached Earth from Andromeda, and this timing could have given yet another motivation for the Magi—now including Liu—to leave immediately for Judea.

All astronomers locate interesting sky positions using asterisms. *Asterism* is the technical term for a group of stars that somehow form a memorable pattern. The brightest stars that could have been used as guides to locate the rising point of Jupiter and Saturn on September 21, 6 B.C., are α-Cassiopeiae (magnitude 2.23) and β-Cassiopeiae (magnitude 2.27), and α- and β- Andromedae (both magnitude 2.06). The prefixes α- and β- mean that these stars were and are the brightest and second brightest stars in a constellation. Thus, "α-Cassiopeiae" means the "brightest star in the constellation Cassiopeia." Using these four stars as rough guide stars would have pointed the eye toward the rising location of Jupiter and Saturn on the autumnal equinox in 6 B.C. in

Babylon, and an observer's eye would have passed directly through M31, as that eye moved from Cassiopeia through Andromeda to the horizon just before the rise of the conjoining Jupiter and Saturn. Thus, if these guide stars were used, an observer would have had a good chance of seeing a supernova in Andromeda just as Jupiter and Saturn were rising. That is, the supernova would first have been seen just as the Sun was setting, which is to say at the acronychal rising of Jupiter and Saturn.

Seeing such a supernova under these conditions would have immediately suggested to an astrologer a connection with a King of the Jews. The supernova would have been seen near the second of the triple conjunction, which was already associated with the Jews, and in Andromeda, which was associated with Palestine, as I described. On the autumnal equinox, M31 transits the zenith at Babylon at 11:12 P.M. (a similar time in Bethlehem), so the Magi would have had no difficulty determining that a supernova passed directly over Bethlehem, since it would have transited near midnight. If Jesus were born on the autumnal equinox, he would have been conceived nine months earlier, on the winter solstice, which as Hughes has pointed out, was thought in ancient times to be December 25.[35] Christian doctrine has always held that life begins at conception, not at birth.[36] For Christians, then, God entered the world at the instant of Jesus' conception, which should be the key date to be celebrated. If Jesus' birth was on the autumnal equinox, then the celebration on December 25 indeed is a celebration of his conception.

A third possibility for Jesus' birth date must be kept in mind. As I shall discuss in more detail in the next chapter, the Christian tradition asserts that Jesus was born without Original Sin. In modern terminology, he was genetically distinct from us: the genes that code for sinful behavior were not present in Jesus. He was what we were intended to be, a "lamb without blemish." The Christian theory of the Atonement rests on the absence of sin in Jesus, original or otherwise. Given the observed fact that sinful behavior is exceedingly common in everyone else in history (excepting possibly Jesus' mother, Mary), the genetic pattern for evil in brain development may be extensive. In view of this possibility, a person who does not possess the genetic pattern for sin may develop more rapidly in the womb, being ready for birth before the normal nine months. If so, then Jesus could have been born in late July or August and kept in his place of birth for a month or so, because his parents, thinking him premature, were afraid to return home.

In this case, a December 25 conception would result in a birth in the Jewish month of Av, the month that ancient Jewish tradition claims the Messiah will be born. We hear in Luke 2:11 that the shepherds were told by an angel "unto you is born *this day* in the city of David a Savior, which is Christ the Lord" (my emphasis). "This day" could have been earlier than the day the Magi first saw the Star. August would have been a month when we would expect shepherds to have been with their flocks. There is no reason why the shepherds and the Magi had to make their homage on the same day, or in the same month.

Supernova remnants have recently been detected in the Andromeda Galaxy,[37] and with improvements in technology, we may expect their number to increase substantially. So supernova remnant observers should look for a supernova remnant in M31 (or its halo) that can be dated to an explosion first visible on Earth 2,000 years ago. Extending the calculation of G. de Vaucouleurs and H. G. Corwin for S Andromedae,[38] the supernova remnant should today have a radius of 6 arc minutes, and if the supernova were of Type Ia, should have iron-rich ejecta, and so may be visible as an absorption nebula in the Fe I resonance line at 3,860 angstroms, as S Andromedae was visible.[39] With substantial improvements in our knowledge of how supernova remnants evolve, it might even become possible to obtain a date of supernova denotation sufficiently precise to distinguish between March 22, 8 B.C., and September 21, 7 B.C. Such dating precision is, of course, impossible today.

The supernova might have been a Type Ic hypernova, a supernova type that is physically a Type II but from a progenitor missing its outer hydrogen envelope. Hypernovae are rare—10^5 supernova occur for each hypernova[40]—but hypernovae are much brighter—they are believed to be the energy sources for gamma ray bursters, which have a top electromagnetic power output of 10^{52} ergs per second[41] (by comparison, the Sun has an electromagnetic power output of 4×10^{33} ergs per second), so the peak power output of a gamma ray burster is more than 10^{18} times greater than that of the Sun. To put it another way, a burster is brighter than 1 million trillion Suns. If hypernovae are like Type II supernovae in having 99 percent of their peak power output in the form of optically invisible neutrinos,[42] then the true peak power would be 10^{21}

times the Sun's output of energy, which means that at peak a hypernova outshines all the stars in the entire visible universe of 10^{20} stars.[43]

I myself find the idea of a hypernova that is barely visible on Earth but that intrinsically (and mainly invisibly) outshines all the stars in the entire visible universe an appropriate star to announce the birth of a carpenter's son who was actually God born of a virgin. Also, in a hypernova, there are three main forms of energy: (1) neutrinos, (2) photons, and (3) gravitational collapse energy. The release of the gravitational energy by the collapse of a stellar core of between one and ten times the mass of the Sun is the source of energy that moves outward as photons and neutrinos.[44] As we shall discuss in Chapter 8, the Son is associated with neutrinos, and as we discussed in Chapter 3, the Final Singularity—God the Father—is associated with gravitational energy. We are told in Genesis that at the Initial Singularity—God the Holy Spirit—there was only light (photons). So in a hypernova we have a nice image of the Trinity. It would be completely appropriate if the Incarnation were announced by a superb astronomical image of the Trinity.

Most expect the central remnant of a hypernova to be a black hole rather than a neutron star, which is the typical central remnant of a Type II supernova. But some have argued for a neutron star.[45] In either case, we might expect a remnant much like the Crab Nebula (M1): a neutron star whose rotational energy excites an emission nebula. The Crab Nebula has an absolute visual magnitude of −3.2, and its pulsar has an absolute visual magnitude of +4.5. Were M1 in M31, the Andromeda Galaxy, the corresponding apparent visual magnitudes would be +20.8 and +28.5, respectively. The sensitivity of the Hale Telescope is +23 visual magnitude, while the Hubble Space Telescope is +28 visual magnitude, as is the Keck I Telescope. The Next General Space Telescope is expected to have sensitivity +31 magnitude. So the Hale Telescope could see an M1 in M31, whereas the Hubble and the Keck would be marginal for detecting the Crab Pulsar in M31, even if it were beaming toward us. Of course, a smaller telescope than the 5-meter Hale will be sufficient for the emission nebula. The 2.4-meter-diameter Hiltner telescope with electronic imaging was adequate for detecting and studying the hypernova remnant MF 83 in the Galaxy M101.[46]

Other meanings can be given to the word *star* in Matthew 2:2. A short list of authors who have suggested other interpretations is Raymond E. Brown; D. W. Hughes; David Clark, John Parkinson, and F. Richard Stephenson; M. R. Molnar; and M. Kidger.[47] Not surprisingly, these yield other interpretations of the Bethlehem event: the triple conjunction I discussed. But if we look for a supernova remnant in M31 at the indicated declination, the literal interpretation becomes uniquely testable. Conversely, finding the predicted supernova remnant in the place indicated would confirm the literal truth of the Magi story of Matthew and would provide evidence that his story of the birth of Jesus was correct.

VII

The Virgin Birth of Jesus

Now the Birth of Jesus Christ was on this wise: When as his mother Mary was espoused to Joseph, before they came together, she was found with child of the Holy Ghost. Then Joseph her husband, being a just man, and not willing to make her a public example, was minded to put her away privately. But while he thought on these things, behold, the angel of the Lord appeared unto him in a dream, saying, Joseph, thou son of David, fear not to take unto thee Mary thy wife: for that which is conceived in her is of the Holy Ghost. And she shall bring forth a son, and thou shall call his name JESUS [Savior]: for he shall save his people from their sins. Now all this was done, that it might be fulfilled which was spoken of the Lord by the prophet saying, Behold, a virgin shall be with child, and shall bring forth a son, and they shall call his name Emmanuel, which being interpreted is, God with us.

MATTHEW 1:18–23 (KJV)

And in the sixth month the angel Gabriel was sent from God unto a city of Galilee, named Nazareth, to a virgin espoused to a man whose name was Joseph, of the house of David; and the virgin's name was Mary. And the angel came in unto her, and said, Hail thou that are full of grace, the Lord is with thee: blessed art thou among women. . . . And the angel said unto her, Fear not, Mary: for thou has found favor with God. And, behold, thou shall conceive in thy womb, and bring forth a son, and shall call his name JESUS. . . . Then said Mary unto the angel, How shall this be, seeing I know not a man? And the angel answered

> and said unto her, the Holy Ghost shall come upon thee, and the power of the Highest shall overshadow thee: therefore also that holy thing which shall be born of thee shall be called the Son of God.
>
> LUKE 1:26–28, 30–31, 34–35 (KJV)

Biblical Interpretation and the Virgin Birth

The verse in which Matthew refers to a prophet foreseeing a virgin conceiving a son is a translation of words of the prophet Isaiah (7:14). For centuries, Christians and Jews have debated whether Matthew correctly translated the Hebrew word in Isaiah. The Hebrew word appearing in the "Hebrew Bible" is *'almah,* which is translated in traditional Christian Bibles as "virgin." However, the usual modern Hebrew word for "virgin" is not *'almah* but *betulah.* The word *'almah* means "young woman" or perhaps "maiden" in modern Hebrew. (By "modern Hebrew" I mean Hebrew as Jews have understood it over the last 1,000 years.) So why do Christian Bibles translate the word in Isaiah 7:14 as "virgin"?

Christians translate the word as "virgin" because Matthew explicitly says it means "virgin." And Matthew says it means "virgin" because he was using the standard Greek translation of his time, the Septuagint, and the Greek word is *parthenos,* which almost always meant "virgin" (but could mean "maiden"). So since Matthew was using a translation, and we have the original Hebrew Bible, shouldn't we accept the meaning of the latter?

Unfortunately, it's not that simple. We don't actually know which version of the Hebrew Bible was being used by the translators-creators of the Septuagint. According to a supposedly contemporary source (the Letter of Aristeas), the translation was begun in the time of the Greek-Egyptian king Ptolemy II Philadelphus, who reigned from 285 to 246 B.C. The translation was made by seventy-two Jewish scholars (six from each of the twelve tribes of Israel) who were sent from Jerusalem at the request of the king for the purpose of translating the Hebrew Bible. The number of scholars gives the name of the translation: *Septuagint* means "seventy," and the Septuagint is often simply denoted by LXX, the Roman numeral for seventy. The modern Hebrew Bible, the Masoretic Text (from the Hebrew word *masoreth,* "tradition"), was begun in the sixth century A.D. and completed in the tenth century. For 600 years it has been the unquestioned version among the Jews. But this final canonical form of the Hebrew Bible goes back earlier: it is a carefully edited version of a canonical Old Testament that was fixed by a synod of Jewish rabbis at the Palestinian town of Jamnia (Jabneh) in the last decade of the first century A.D. The five books of Moses (the Torah)—Genesis, Exodus, Leviticus, Numbers, and Deuteronomy—possibly were chosen as early as 622 B.C.[1] The rest of the Old Testament was probably fixed by 300 B.C.,[2] which is about the time the Septuagint was written.

But even though the oldest versions of the Old Testament that we have today (the Dead Sea Scrolls date from the second and first centuries B.C. and are the oldest)[3] agree very closely with the Masoretic Text, there are some differences. The Samaritans, a community of Jews who claim to be descended from the Palestinian Jews who were not deported to Babylon when Israel was conquered by the Assyrian Empire in 722 B.C. (the Talmud hints that instead they are descended from the peoples the Assyrians brought in to replace the deported Jews), have their own version of the Torah. The Samaritans claim their version goes back to Abishua, the great-grandson of Aaron. The Samaritans' version of the Torah is unquestionably ancient, and most important, it has been copied and recopied independently of the Masoretic Text. It differs from the Masoretic Text in some 6,000 places, but in some 1,900 of these it agrees with the Septuagint.[4]

So perhaps the Septuagint used by Matthew, and not the Jewish Masoretic Text, is actually closer to the original version of the Old Testament. After all, this translation was made just shortly after the books of the Old Testament were fixed by Jewish tradition. And it should be kept in mind that when the Masoretic Text was selected, it was in the Christian era, just as Jews and Christians began to argue in earnest about the correct interpretation of biblical passages. It is remotely possible that the rabbis of the late first century A.D. chose a version of Isaiah that used *'almah* rather than *betulah,* though I want to emphasize there is no evidence whatsoever that such a version of Isaiah ever existed. Among the Dead Sea Scrolls, for example, there is a complete Isaiah, and *'almah* is used in Isaiah 7:14. But my point is, we don't know. Whereas we do know that the Septuagint translated whatever word was there as "virgin."

A more likely possibility is that the meaning of the word *'almah* changed over the four centuries between the translation of the Septuagint and the setting of the Masoretic Text. Think of the English word *maiden*. The general meaning of *maiden* is "young girl," but in fact "virgin" is a secondary meaning. After all, a "young girl" is usually a "virgin," at least if the girl is sufficiently young, and this connection is the reason *maiden* took on the secondary meaning of "virgin." Which of the two meanings of *maiden* is the correct meaning is to be decided by the context.

The context in which a word is used can change the meaning of a word. Take the English word *weird*. It originally meant "fate." Centuries ago, an English speaker could say "It was his weird to die young" to mean "It was his fate to die young." But William Shakespeare wrote a play, *Macbeth,* in which three strange witches foretold the fate of the central character. The fact that there were three witches shows that Shakespeare wanted his audience to think of these witches as the three Norns, the three pagan goddesses of fate, whose names were That Which Was, That Which Is, and That Which Shall Be. Past, Present, and Future. So Shakespeare referred to his witches as the "weird sisters," indeed another title of the three Norns. Shakespeare was far more widely read than pagan myth, and over time the original meaning of the "weird" sisters was forgotten, but instead the connection with the "strangeness" of Macbeth's interaction with the three witches was remembered. So *weird* took on the now primary meaning of "very strange," while the original primary meaning of "fate" was forgotten (but still recorded in the most complete dictionaries).

There is some evidence that in the time of Isaiah's writing and earlier, *'almah* carried the primary meaning of "virgin." In the Hebrew Bible, the word *'almah* appears in the singular three times: in Genesis 24:43, where is it used in reference to Rebecca; in Exodus 2:8, where it is used in reference to Moses' sister; and in Proverbs 30:19, where it appears in the phrase "the way of a man with an *'almah"* ("maid" in the King James Version).[5] In the cases of Rebecca and Moses' sister, it is clear from the context that they were both unmarried at the time, and presumably virgins. My own interpretation of Proverbs 30:19 would suggest "virgin" is appropriate here too. In Isaiah 7:14 the form of the word *'almah* includes an article, so a more precise translation is not "a virgin/maiden" but rather *"the* virgin." This of course fits in nicely with the traditional Christian interpretation, but Jewish rabbis considered that the context required the woman to be a contemporary of Isaiah, and they debated over whom "*the* young woman" could be.[6] According to Christians, *the* virgin could be none other than Mary, the Mother of Jesus.

And of course the Jewish rabbis insist that *'almah* definitely means "young woman" rather than "virgin." Since the Septuagint says otherwise, I suggest that the word changed from a primary meaning of "virgin" in 300 B.C. to something closer to "young woman" around A.D. 90, when the major rabbis began to engage Christianity in a serious way. For example, in Chapter 57 of Justin Martyr's *Dialogue with Trypho, a Jew,* written about 130, Trypho indeed argues that the proper translation of the Isaiah passage is not "virgin" but "young woman." If there were any ambiguity in the meaning of *'almah* in A.D. 90, all Jewish rabbis would have had a strong unconscious motivation to select the meaning that would make nonsense of the Christian claim. Justin Martyr makes precisely this rebuttal in Chapter 71: the Jews, he claims, have changed the correct translation of the Septuagint. Justin Martyr was certainly correct in pointing out that the Septuagint contained books of the Old Testament that the rabbis rejected, for example, the books of the Roman Catholic Old Testament called the Apocrypha. Some of these are 1 Esdras, the Wisdom of Solomon, Ben Sira (Ecclesiasticus), Judith, Tobit, Baruch, and the two books of the Maccabees.[7]

For me, the conclusive proof that *'almah* can still, even in the modern Hebrew Bible, mean "virgin" in the right context is the fact that two major Jewish biblical commentators of the Middle Ages said so. The most famous of all Jewish commentators on the Hebrew Bible was Rabbi Shlomo Itzhaki, better known by his acronym, Rashi, who lived between 1040 and 1105, mostly in the French city of Troyes, a location we shall encounter again when we discuss the legend of the Holy Grail. In the Song of Solomon, known to Jews as the Song of Songs, the word *'almah* appears in verse 1:3 in the plural: *'alamot.* According to Rashi, in this verse, the word means not "young women" but "virgins."[8] (The King James Version translates the word as "virgins.") Needless to say, in his commentary on Isaiah, Rashi follows the standard Jewish tradition after A.D. 90 that *'almah* in Isaiah 7:14 means "the young woman." Another major Jewish rabbi, Gersonides (Levi ben Gershom, who also is known by the acronym Ralbag), who lived from 1288 to 1344, mainly in the south of France, made it very clear in his commentary on the Song of Songs that *'alamot* in verse 1:3 means "virgins": "And therefore *do the maidens* love him—these maidens being the young girls who had not been with a man."[9]

If the two greatest medieval Jewish commentators on the Hebrew Bible claim that *'almah* can in the proper context mean "virgin," then I claim it can. This does not settle the question of whether in the context of Isaiah 7:14 *'almah* is appropriately translated "virgin." Both Christians and Jews agree that the true meaning of a prophetic passage can often not be understood until the prophecy comes to pass. So I shall now investigate whether the Christian interpretation of Isaiah 7:14 has indeed come to pass. Has a Son been born of a Virgin?

The great German theologian Wolfhart Pannenberg has rejected the Virgin Birth.[10] However, Pannenberg has a solid Trinitarian reason; he believes,

> In its context, the legend of Jesus' virgin birth stands in an irreconcilable contradiction to the Christology of the incarnation of the preexistent Son of God found in Paul and John. For, according to this legend, Jesus first *became* [Pannenberg's emphasis] God's Son through Mary's conception. According to Paul and John, on the contrary, the Son of God was already preexistent and then as a preexistent being bound himself to the man Jesus.[11]

The problem with the clause "and then as a preexistent being bound himself to the man Jesus" is that it suggests the Adoptionist heresy. Indeed, as I argued in Chapter 4, the Son is preexistent, has existed since "before" time began. Thus, as Pannenberg emphasizes, Jesus did not become God's Son through Mary's conception. But if Jesus' birth were not a virgin birth, then Jesus would have to have had a human biological father. If we assume Mary was an honorable woman, this biological father would have to have been Joseph. In which case Jesus would have had two fathers, Joseph and God. If a man has two fathers, the father who is not the biological father is called the "adopted" father. This would still be true even if we were to imagine that the Son were to unite with the man Jesus at the very instant of conception. As Pannenberg is aware, the Virgin Birth has been used since Luke to establish the Trinitarian Dogma.[12] When Trinitarianism loses authority, the heresy of Adoptionism appears. Adoptionism is intimately connected with the Arian heresy. The Adoptionist heresy claims that Jesus the man was not God the Father's "natural" son but was instead an ordinary man "adopted" by God (in most versions of this heresy, at the time of Jesus' baptism by his cousin John the Baptist).

Matthew and Luke claim that Mary was a virgin when Jesus was conceived. Pannenberg considers these passages to be "a legendary tradition that has been incorporated by Luke in his gospel and has been alluded to in Matthew."[13] Pannenberg also argues that the literary form of the Lucan story indicates that it is a legend.[14] Other experts on "form criticism" (a technique of biblical analysis that attempts to interpret the meaning of the words), Raymond Brown, Rene Laurentin, and Manuel Miguens, all disagree with Pannenberg, as do the literary experts C. S. Lewis and Dorothy Sayers.[15]

Form criticism has come under attack by biblical scholars themselves in recent years, as they realized that the conclusions reached depended more on the philosophical and theological presuppositions of the form critics than on the actual texts.[16] The "philosophical" presuppositions were clearly stated by the leading form critic of the 1930s, Rudolf Bultmann: "Myths [like the Virgin Birth] are difficult to believe in these days of electric lights."[17] In other words, the "philosophical" presuppositions were not biblical, or even theological, but physical; form critics based their rejection of the miracles of the Bible on their amateur knowledge of physics! The question of the truth of the Bible stories is a question of physics, and I am a much better physicist than any form critic.

However, I think both sides of the literary argument have ignored facts from the natural sciences that bear on the question of the Virgin Birth of Jesus. First of all, there is the dating of Matthew and Luke. The consensus date range today is A.D. 75–90, after the destruction of Jerusalem in 70, too long after Joseph's and Mary's lifetimes for us to expect the Gospel authors to have had firsthand input from the only ones who would have known. John Robinson's redating of Matthew and Luke to A.D. 40–60[18]—well within Joseph's and Mary's lifetimes—has not been widely accepted. But Acts ends with Paul awaiting trial in Rome, strongly suggesting that Luke completed Acts (and his Gospel, which obviously precedes Acts of the Apostles) before Paul's trial, conviction, and execution. Biblical scholars have nevertheless rejected this implication because they believe there are references in Luke to the destruction of Jerusalem.[19]

On this question I can apply my experience as an astrophysicist. We astronomers are always asked about claims of astrologers to predict the future, and we have developed criteria to evaluate such claims. We have always found that before the event, an astrologer's "prediction" is very vague, and after the event his or her "prediction" becomes very precise—naturally, because after the event the astrologer knows when the event occurred and all the details of the event. Before 9/11, astrologers "predicted" that terrorists were going to attack some American target eventually. Didn't we all know this before 9/11? After 9/11, astrologers claimed to have "predicted" that the stars showed that jets were going to crash into New York buildings in September 2001. The "prediction" of the destruction of Jerusalem in Luke has all the hallmarks of a prediction made before the fact. That is, Luke's "predictions" are more consistent with being written before A.D. 70 than after. At the very least, we cannot use these passages to conclude that Luke and Matthew were written after 70. So the other evidence that Luke was written before the death of Paul retains its full force, and Robinson's earlier date is more believable than the consensus date.

I might also add that I have become very suspicious of any "consensus" claim. I was once myself part of a scientific "consensus," a consensus in the 1970s which claimed that the universe was not accelerating. This was indeed the consensus of cosmologists in the 1970s. We now have very strong observational evidence that that consensus was wrong. The universe is indeed accelerating. We members of the "scientific consensus" appealed to "the consensus" because we did not have good evidence for our position. I'm older but, I hope, wiser. I want to see evidence, not hear about a "consensus." I'm inclined now to regard the word *consensus* as a synonym for "wrong." As I discussed in Chapter 3, had I not been blinded by the consensus view in cosmology in 1994, I would have predicted the acceleration of the universe. I even then knew that known physics required an acceleration.

It is often claimed that Mark, John, and Paul never mention the Virgin Birth. It is certainly true that the word *virgin* is never used in the New Testament in connection with Jesus' birth except in the nativity narratives in Matthew and Luke. However, there is an astounding omission in all modern discussions of the Virgin Birth: no biblical scholar ever gives an analysis of the ancient theory of human reproduction. This is an exceedingly serious defect in all modern discussions of the Virgin Birth, because it is possible that an ancient reader, who naturally would have known the ancient theory, would have recognized a reference to the Virgin Birth in a biblical passage where a modern reader would not.

I shall argue that this is in fact the case. There are numerous references to the Virgin Birth in Mark, in John, and in the Pauline letters. Only moderns do not see these references, because we are no longer familiar with the reproductive theory of the ancient world. It is no coincidence that doubts about the Virgin Birth first became widespread in the nineteenth century, just when the modern theory of human reproduction was developed.

There were actually at least two major ancient theories of human reproduction, which I can only outline here, and which for our purposes are essentially equivalent. For more extensive discussions of these theories, see Peter Bowler, Michael Boylen, F. J. Cole, Anthony Preus, and especially Julia Stonehouse.[20] It was of course known in the ancient world that children resulted when a man injected a fluid material into the womb of a woman. The question was, What exactly was in the fluid that caused the child to be generated? Both ancient theories rejected our modern notion that the genetic material came equally from the mother and the father. The ancients believed that the genetic "material" came from the father alone. If a child resembled the mother, this was the result of environmental effects. The child was generated in basic essence by the paternal "substance." This is why descent in the ancient world was almost always patrilineal and the child was considered the property of the father: in the ancients' eyes, the child *was* related only to the father. There is a passage in the *Eumenides* by the ancient Greek playwright Aeschylus (lines 657–666) in which the god Apollo defends Orestes from the charge of matricide by arguing that the mother is only a nurse to a child; the genetic material comes solely from the father.

There are numerous biblical passages in which the male material is referred to as "seed," and in the ancient world, this word meant "fertilized seed." Seed, in other words, which has all of the genetic material. The descendants of King David were referred to as "the seed of David," again implying that the genetic material comes entirely from the male. As is well known, this convention is general: "seed of a man" is equivalent to "descent from that man." But this was no mere convention. The language expressed a definite theory of how reproduction operates. Theories differed concerning how the genetic information was coded in the male material. But they all agreed that the genetic material comes only from the father.

Aristotle's theory of genetic coding was the dominant such theory in the Greek-speaking world, which was the world of Paul and of the Gospel writers. According to Aristotle, the male was the efficient and formal cause of the child, while the female was the material cause. That is, the male started the process (efficient cause) and provided the program-genetic code (formal cause). Aristotle believed that the program of life was imposed on the menses of the woman. The woman provided just the material, just as the soil provides the material for a seed to become a plant. Thus, a child was begotten by a man and merely borne by a woman. Galen (A.D. 129–199) advanced a "two-seed" theory—one seed from the male and one from the female—but even in this theory the genetic material came entirely from the male. Galen's theory could be viewed as somewhat similar to our sperm and egg theory, but with all the genetic information in the sperm. A major reason why virginity in women before marriage was so important in the ancient world is that the seed-sowing picture was accepted literally. If a woman had sexual intercourse with a man other than her future husband, then the lover would permanently contaminate her womb with his seed, as sowing a field with the seeds of weeds would contaminate the field, at least for a very long time.

Aristotle's form-imposed-on-matter reproduction theory is expressed in John 1:14—"And the Word was made flesh . . . as of the only begotten of the Father." In other words, God (the Holy Spirit) imposed His form on the matter in the womb of Mary. And only if Mary were a virgin before this was done can we humans be assured that Jesus really is God's Son. Only if Mary were a virgin could Jesus legitimately be called God's Son. In reference to John's Gospel, we must always remember that the first non-Apostle Church Father to use the explicit phrase "Virgin Birth" was Ignatius of Antioch (d.c. 110), and by tradition he was an auditor of St. John; *auditor* means that he actually heard St. John speak.

Notice that, in the ancient theory of reproduction, Jesus' preexistence holds automatically if Mary were a virgin. The Form of Jesus comes entirely from God, who naturally existed before the universe was created. Thus, Jesus necessarily was also preexistent, at least in Form. His unimportant material makeup, which came from Mary, did not come into existence before Mary did. But then most of the material making up Jesus came from the food he ate. Both moderns and ancients are agreed that the human form is imposed on the eaten food and not vice versa. Pannenberg and earlier German theologians such as Rudolf Bultmann are wrong about the Virgin Birth being inconsistent with preexistence.[21]

Mark's Gospel opens with "The beginning of the gospel of Jesus Christ, the Son of God." The evidence that Mark intended the phrase "Son of God" to signify the implantation of the Form of God into Mary's virgin womb is in Mark 14:61–64: "Again the high priest asked him [Jesus] . . . 'Art thou the Christ, the Son of the Blessed?' And Jesus said 'I am. . . .' Then the high priest rent his clothes, and saith, 'What need we any further witnesses? Ye have heard the blasphemy.' " But if Jesus were not asserting himself to be the Son of God in the sense that everyone at the time would interpret this claim, why would the high priest consider the claim blasphemous? Conversely, the fact that the claim was considered blasphemous is a strong indication that the high priest was interpreting—and most important, Mark himself was interpreting and intended his readers to interpret—the claim as Jesus' birth from a virgin.

In his letters, Paul refers to Jesus as God's Son forty-one times. He uses different versions of this expression, as is usual in any language. But once one realizes that Paul presupposes the ancient theory of reproduction, one realizes that he presupposes the Virgin Birth. For example, Romans 1:3–4: "Concerning His Son Jesus Christ our Lord, which was made of the seed of David according to the flesh; and declared to be the Son of God with power . . . by the resurrection from the dead." Notice the use of the Galen theory of human descent, as is made clear by Romans 8:3: "God sending His own Son in the likeness of sinful flesh," and of course by Galatians 4:4. The German biblical scholar Adolf Harnack and more recent biblical scholars are quite wrong about the Virgin Birth and the New Testament.[22] Once the ancient theories of human reproduction are understood, references to the Virgin Birth can be seen throughout the New Testament. But it is absolutely essential to understand the ancient reproduction theory to see this. A correct translation from a language has to take into account the scientific presuppositions of the time. And of course, the meaning of a word depends on the date it was written. When I was a boy, calling a man "gay" meant that he was a happy person.

A third mechanism of reproduction—fertilization of eggs by the wind—was occasionally discussed in the ancient world.[23] Augustine and Origen wrote about this "phenomenon" (I put it in quotation marks, because it did not actually exist). It was generally accepted in the ancient world because it fit so well with Aristotle's theory of forms. Recall that, in both Hebrew and Greek, the words for "spirit" and "air" (or "wind") are almost the same word, which is why passages in the Bible can be translated either as "spirit" or "air/wind." Thus, Genesis 1:2 is sometimes translated as "And the Spirit of God moved upon the face of the waters," and sometimes it translated as "And a mighty wind blew over the face of the waters."

But these arguments based on the literary structure of the New Testament have always seemed to me to be beside the point. A much stronger argument against the Virgin Birth has always been the one put forward by the atheists: if Jesus was virginally conceived, the only one who could have known this was Mary, and why should we believe her? She might not even have known herself. Maybe she was raped while drunk. Furthermore, if there was anything irregular about Jesus' conception, suggesting His father was some man besides Joseph—and we are told in Matthew 1:19 that Joseph himself at first believed this—then Mary would have an enormous motivation to lie. The Jewish Talmud makes the claim that Mary became pregnant by another man before her marriage to Joseph.[24] But the Church Fathers have universally affirmed the Virgin Birth because, in spite of a lack of supporting evidence, it has seemed essential to the Trinitarian Dogma.

In addition, once one omits one central part of the Christian tradition on the basis that it sounds "legendary," where does one stop? Arian heretics such as Isaac Newton—and of course atheists—use similar arguments to justify rejecting the Trinitarian passages in the Gospels and in the Pauline letters as nothing but similar legendary traditions. Newton believed that the great theologian and bishop of Alexandria Athanasius (293–373) was the culprit responsible for imposing the Trinitarian "legend" on the Christian world. Indeed, Athanasius was a very important theologian opposing Arianism. So important is Athanasius's reputation in establishing the traditional Christian belief in the Trinity that a short summary of this doctrine is attributed to him. I've included this Athanasian Creed in the appendix.

Other heretics regard Mark without the Resurrection passages as the only valid Gospel and claim that the Resurrection passages in Matthew, Luke, and John are merely later inventions. These people point out that legends of a God who dies only to be resurrected are quite common in the ancient world. Indeed, they were common, but the Gospel accounts of the Risen Jesus have in my judgment (and Pannenberg's and that of most other scholars who have studied the matter with open minds) a ring of reality unlike these myths. Similarly, the accounts of the Virgin Birth in Matthew and Luke have the ring of reality, unlike the equally common ancient myths of the conception of a god born of copulation between a god and a human female. Matthew and Luke describe the Virgin Birth as the result of the action of the Holy Spirit, *not* as the result of intercourse between God the Father and Mary. I propose that Christians should first try to develop a theology based on the Gospels and the Pauline letters in the form given to us. We should assume, at least initially, that there are no legends or human inventions involving important Christian dogmas in the New Testament. This is a theological version of my scientific approach as described earlier: no firmly tested physical law is to be set aside without experimental justification.

Scientific Explanations of the Virgin Birth

I shall now describe a simple mechanism, completely consistent with known physical law, whereby a virgin birth can occur via the action of the Father through the Holy Spirit. In this mechanism, the mind of the virginally conceived Jesus would be in resonance with, and in complete harmony with, the Son from the instant of formation of the mind in the nervous tissue of the embryo. He would be completely human, with the rational mind of a human, but nevertheless be the Son. Furthermore, I shall show that if the Virgin Birth occurred in the manner I propose, the Virgin Birth hypothesis can be verified by direct experiment. The Virgin Birth would no longer rest on Mary's word alone. We would be able to show directly, without reference to human testimony, that Luke and Matthew merely reported the facts as related to them by a completely truthful Mary. A direct experimental confirmation of the Virgin Birth would also support the claim that Matthew and Luke were just reporting the facts when they described the Risen Jesus.

We first have to understand how a virgin birth of a human male can be accomplished using only known molecular biological mechanisms. There is now an enormous scientific literature on virgin birth in vertebrates.[25] Virgin births—more often called *parthenogenesis* in the scientific literature, after the Greek word for "virgin birth"—have been extensively studied in Caucasian rock lizards and also in turkeys.[26] There is one strain of turkeys in which more than 40 percent of all births are virgin births. Often in these turkeys a haploid egg cell begins to divide without being fertilized by a sperm cell. If at some point early in the cell division process, the chromosomes duplicate so that a diploid cell is formed, a normal turkey is born. This parthenogenetic turkey is always a male, because birds use a WZ sex determination system. In a WZ system, a male results if the two sex chromosomes are the same. Thus, a male bird has two Z chromosomes, and a female has the mixed sex chromosomes WZ. The combination of two W chromosomes is always lethal. We humans, as do all mammals, use an XY sex determination system. A female results if the sex chromosomes are identical XX. A male results if the sex chromosomes are different: XY is the male. The combination of two Y chromosomes is always lethal in humans.

Reptiles use yet another sex determination system. They do not have separate sex chromosomes. Instead, all their chromosomes are *autosomes,* which means that the chromosomes occur in pairs that look the same under the microscope. Sex in reptiles seems not to be genetic, coded in the DNA, but environmental: a reptile is a male or female depending on the average temperature experienced by the egg during development. Warm-blooded animals such as mammals and birds obviously need a different, nontemperature-based system, and we have it. But birds and mammals, including humans, are nevertheless similar to reptiles in that virtually all that the Y chromosome does is instruct male genes on other chromosomes to activate. The Y chromosome itself has no true sex determination genes.

Virgin birth in a snake (python), starting from a diploid cell, has recently been confirmed using DNA analysis, which showed that the daughter snakes had exactly the same DNA as the mother.[27] A haploid virgin birth would yield daughters with only half the DNA as in the mother but with no DNA that is not also in the mother.

It is easy to induce a human oocyte (egg cell) to begin cell division without first being fertilized by a sperm.[28] The oocytes thus induced can be either haploid or diploid. This human oocyte cell division is so easy to induce in the laboratory that many researchers in this field have suggested that virgin births may be quite common in humans, perhaps as common as identical twins, which on the average occur 1 out of every 300 births.[29] This conjecture could be easily tested. One would merely conduct a DNA identity test on female children who are observed to closely resemble their mothers. (Almost all virgin birth children would be expected to be females, since a female has only two X chromosomes, even though the essential genes to create a male are available in every human female, on other, non-X chromosomes. The extremely rare exception I shall discuss shortly.) Unfortunately, up to the present, no such investigation has been carried out, possibly because of ethical objections. Or it could simply be that most people, even doctors, have not realized how easy it is to test for a virgin birth in a human. And there is a mental barrier to even considering that a virgin birth can occur in a human.

There are many cases in the literature in which a woman claims that she has conceived without having sexual intercourse. More precisely, the woman claims that she merely "fooled around" with her boyfriend but did not allow penetration by his male organ (like the interaction between President Bill Clinton and his aide Monica Lewinsky). There was nevertheless semen on the woman's body, and all doctors hearing this story surmise that male sperm somehow entered the woman and fertilized one of her eggs. However, the sort of sexual stimulation that occurs in these cases is very similar to the stimulation two female lizards give each other in order to induce a virgin birth in each. So it is at least *possible* that the stimulation can on rare occasions induce a virgin birth in a human female.

If the child of a woman who has conceived without sexual intercourse is a female, a twin zygosity test should be routinely performed on mother and daughter to see if they are genetically identical. The now standard test for monozygotic twins (that is, identical twins) uses DNA at five distinct locations in the genome, and the DNA at each of these locations varies enormously from one person to the next.[30] If all the DNA markers are the same, then the odds are about 100 to 1 in favor of the two individuals being twins. If the two individuals were mother and daughter, the odds would be 100 to 1 in favor of a virgin birth. One caveat concerning this test should be pointed out. All probabilities are conditional probabilities, which is to say that the probability, which estimates human ignorance, takes into account all that we actually know about a given situation. In the twin zygosity test, the test assumes that the two people tested are known to be twin siblings of the same gender. But in using the twin zygosity test to test for a virgin birth, we are actually assuming that the two people being tested are mother and daughter. This means that the conditional probability will be different. But the two probabilities are not too different. If we denote by the symbol p_i the frequency in the general population of the ith variant of a gene (the technical term for "variant of a gene" is *allele*), then it can be shown that the probability that two (untyped) siblings have the same allele of a gene is

$$(1/4)(1 + 2\,(\Sigma\, p_i^2) + 2(\Sigma\, p_i^2)^2 - \Sigma\, p_i^4)$$

where the upper case Greek letter Σ just means "sum over all frequency numbers p_i."[31] As the number of gene variants—the number of alleles—increases, the individual frequencies p_i get smaller and smaller, and so the probability that two untyped siblings share the same allele approaches ¼. This is exactly what we would expect, since two siblings on the average share one-quarter of their genes. However, the probability that a parent and child (both untyped) share the same allele is

$$\Sigma\, p_i^2$$

which approaches 0 as the number of alleles of the genes gets very large.[32] This formula is very counterintuitive, since a parent and child necessarily share half of their genes. However, a moment's thought shows the formula makes sense. Siblings both have the same two parents, and so there are many more ways of having the two alleles the same. In the case of parent and child, one parent fixes the gene that the other parent has to provide. As the number of alleles increases, the likelihood that the other parent will provide exactly the same gene as the parent being tested goes to 0.

This means that the odds that a mother-daughter pair will test positive for identity if they take a twin zygosity test are actually greater than the odds for two siblings. More like 1,000 to 1, rather than the odds of 100 to 1 that the company will quote you if you take the test. But let us be conservative and accept the company's odds.

The odds of 100 to 1 are not sufficient to provide convincing evidence of a virgin birth, but if the twin zygosity test on a mother-daughter pair is positive, a more extensive test should be carried out. A custom-designed DNA twin zygosity test using not five but twenty-five (microsatellite) markers can be performed by any major DNA testing service,[33] and if all of these markers for mother and daughter agree, the odds in favor of a virgin birth would be 100^5 (= 10 billion) to 1. This huge number means that the custom-designed test would establish an actual human virgin birth conclusively. The cost of the custom-designed twin zygosity test should be no more than ten times the cost of the standard test, or about $2,000. These tests are for normal twin zygosity, which in the case of a virgin birth means that they will give a standard positive only if the virgin birth is diploid. If it is haploid, the test will show a doubling of the DNA at the sites where the daughter's DNA is the same as the mother's, and an absence of DNA at the other sites. The standard test would show half the variability in the daughter as would be in a normal person, and in the case that the test is negative, the lab should be asked if this reduced variability is present.

To perform the standard (and custom-designed) twin zygosity test, one takes DNA samples via cotton swabs of mother and daughter, and mails the samples to the laboratory. The test is painless and noninvasive. The people being tested can take the samples themselves. Here is a list of five recognized DNA laboratories that can perform the standard twin zygosity test, the quoted 2005 cost of the standard test, and their web addresses.

Gene Tree DNA Test Center $195
http://www.genetree.com/product/twin-dna-testing.asp
Proactive Genetics* $140 + $10 shipping
http://www.proactivegenetics.com/fees.dna
Affiliated Genetics $120 + $10 shipping
http://www.affiliatedgenetics.com/FSTT.htm
Beta Paternity $195
http://www.betagenetics.com/twin-zygosity.html
DNA Diagnostics Center* $400 + $70 collection fee
http://www.dnacenter.com/testing-cost.html

With human oocytes, again for ethical reasons, no attempt has yet been made to implant these virginally conceived fetuses into wombs. An attempt was made to complete a virgin birth in a marmoset monkey by Vivienne Marshall and her group in 1998, but it was unsuccessful.[34] Because of a peculiarity in a cell structure, the centromeres of primate cells (they are inherited from the father), I myself suspect a primate virgin birth could result only if a diploid oocyte started to divide.[35] Of course,

---

*Will accept samples mailed from virtually any country on Earth. Shipping fees apply to U.S. addresses only. Collection fees, which are higher than shipping, apply if legal custody of the DNA sample is desired. The author receives no money from any of these companies and cannot guarantee the quality of their work. Caveat emptor!

in every virgin birth, all the genetic information has to be already present in the mother. There are at least three ways to generate a male human being from genetic information that comes entirely from the mother. I shall discuss the simplest first and in detail, then briefly summarize two others that have been proposed.[36]

I propose that Jesus was a special type of XX male, a type that is quite rare in humans but extensively studied.[37] Approximately 1 out of every 20,000 human males is an XX male. Such males are normal in behavior and intelligence but have smaller teeth, shorter stature, and smaller testes than normal males. They are usually identified as XX males because they cannot have children and ask doctors to cure the infertility. Normal males are XY, but there are only twenty-eight genes on the Y chromosome, as opposed to thousands on the X chromosome. Of these twenty-eight genes, fifteen are unique to the Y chromosome and thirteen have counterparts on the X chromosome.[38] The genes with counterparts on both the X and the Y chromosomes are called *homologous genes*. An XX male results when a single key gene for maleness on the Y chromosome (the SRY gene) is inserted into an X chromosome. One possibility is that *all* (or at least many) of the Y chromosome genes were inserted into one of Mary's X chromosomes and that, in her, one of the standard mechanisms used to turn off genes was active on these inserted Y genes. (There is an RNA process that can turn off an entire X chromosome. This is the most elegant turnoff mechanism.) Jesus would then have resulted when one of Mary's egg cells started to divide before it became haploid and with the Y genes activated (and, of course, with the extra X genes deactivated). If a sample of Jesus' blood and/or flesh could be obtained, my proposal could easily be tested by carrying out two distinct DNA tests for sex: (1) test for the Y genes and (2) test for two alleles (different gene forms) of X chromosome genes. In other words, a male born of a virgin would have two X chromosome genes for each of its counterpart Y genes. Normal males would have only one X chromosome gene for each Y counterpart gene. This pairing would apply to each of the thirteen genes on the Y chromosome that has an X counterpart.

Such a virgin birth would be improbable. If the measured probability that a single Y gene is inserted into an X chromosome is 1 in 20,000, then the probability that all Y genes are inserted into an X chromosome is 1/20,000 raised to the 28th power, the power corresponding to the number of Y genes. (Assuming that the insertion of each Y gene has equal probability and that these insertions are independent.) There have been only about 100 billion humans born since behaviorally modern *Homo sapiens* evolved, between 55,000 and 80,000 years ago.[39] The number of humans who have ever lived is roughly computed as follows. In the first 60,000 years of modern human existence, there were roughly 10 million humans living worldwide, with complete replacement every generation, about every 30 years. With 2,000 generations in 60,000 years, this means 20 billion people lived in this period. Over the next 6,000 years, humans had agriculture, which allowed the support of a population of roughly 300 million. With 200 generations in 6,000 years, this means that 30 billion people lived in this period. Finally we come to the modern period, essentially the period of the people now living. There are now 6 billion people in the world. Adding all these numbers gives about 60 billion as the total number of people who have ever lived.

Thus, the virgin birth of such an XX male would be unique in human history even if there were only two such Y genes inserted into an X chromosome. (I assume an upper bound to the rate of virgin birth is 1/300. Then the probability of a virgin birth of a male with 2 Y genes is 1/[300][20,000][20,000] = 1/120 billion.) But, as in the case of the Resurrection, if such an event *had* to occur for the universe to evolve into the Omega Point, then the Virgin Birth probability would become 1; that is, certain to occur. In other words, it would be a miracle!

A far more probable Virgin Birth would be for the SRY gene alone to be inserted into an X chromosome of Mary, with the Virgin Birth resulting from either a haploid sex cell undergoing chromosome duplication or resulting from a diploid cell. In the haploid case, there would be an SRY gene inserted in each normal X chromosome. In the diploid case, there would be one SRY gene for every two normal X chromosomes. Both genomes are distinguishable by standard DNA tests from normal males. A normal male would have the normal collection of additional Y genes, whereas an XX male with only the SRY would be missing these genes. The standard DNA test today looks for many Y genes. (The sex determination test that was standard in the mid-1990s, invented by Lucia Casarino and others, searched only for the AMEL-X and AMEL-Y genes.)[40] So the standard DNA sex determination test would be able to distinguish all the various ways an XX male could arise via a virgin birth.

There is a final possibility that we must be aware of when considering the genetic signature of a male virgin birth. The SRY gene does not itself generate the male organs, as I mentioned. Rather, it induces other genes located on the autosomes to generate those organs.[41] This raises the possibility that the SRY is itself not necessary. And in fact some XX males have recently been studied in which the SRY gene appears to be absent.[42] This case can also be distinguished from a normal XX female and a normal XY male by today's standard DNA test. Once again, the male genes would be present in the DNA but in a different ratio than in a normal male.

The observed acceleration of the universe provides a possible reason why the Virgin Birth and Resurrection necessarily had to occur if the universe is to evolve into the Omega Point. If the acceleration is to stop eventually and be converted into a deceleration and universal collapse, then our descendants must expand into the universe and annihilate baryons via the electroweak tunneling process. We do not know how to do this. We know only that this process is allowed according to the Standard Model and must have operated in the early universe. On the basis of the Standard Model alone, we have no indication of how to annihilate baryons in a practical way. But if the universe is to evolve into the Omega Point, then there must be a practical, small-scale method of annihilating baryons to provide energy before the recollapse of the universe provides gravitational energy, and to provide efficient relativistic rockets, as discussed in Chapter 3.

Suppose the Son became incarnate to provide us this information. Notice that He can do so only by simultaneously providing us with the knowledge that we ourselves one day will be resurrected with bodies in all essentials like the body Jesus had after His Resurrection. Also, we can obtain the necessary information only by believing in Him, believing that He is God, and believing that He rose from the dead. Without such a belief, no one would investigate Jesus for clues of constructing a practical device for annihilating baryons. If He provides us with the essential hints for how to construct such a device, He saves the entire world. Literally, He saves the entire universe from destruction (violation of unitarity when black holes evaporate). "For God sent not his Son into the world to condemn the world, but that the world through him might be saved" (John 3:17). Traditionally, the word *world* has been interpreted to mean "humankind," but with my proposal for the reason for the Incarnation and Virgin Birth, *world* can be interpreted literally.

I further propose that the Virgin Birth was necessary so that Jesus would have a unique body type that could, out of its own nature, generate electroweak quantum tunneling. On this proposal, Mary, who had the same genes as Jesus, should have been able to accomplish this too, but it is conceivable that only with the male genes activated could this process be fully carried out. (Perhaps Mary was capable of the Assumption but not the Resurrection.) This would solve the problem of why Jesus was a Son and not a Daughter. Why not a Daughter? has always been a problem for Christian theology, though rarely discussed because there was no way of answering it. The theory of the Resurrection and Assumption that I will develop in detail in Chapter 8 may provide one.

If Jesus were an XX male conceived of a virgin in the more complex way just described, with all the genes in the Y chromosome present, we could resolve the two well-known inconsistencies between the genealogies provided by Matthew and Luke.[43] The first inconsistency is the fact that descent is listed in the line of Joseph, who as Matthew and Luke both assert, was not Jesus' biological father. This inconsistency is usually resolved by pointing out that if Joseph acknowledged Jesus as his son (which he did), then according to the Jewish law of the time, Jesus would be considered Joseph's son, and hence "descended" from David if Joseph were so descended.[44] Luke seemed to point out this double meaning of *descent* when he wrote, "And Jesus . . . being (*as was supposed*), the son of Joseph, which was the son of Heli" (Luke 3:23) (my emphasis).

The second inconsistency lies in the fact that the two lists are different, beginning with the father of Joseph (Jacob according to Matthew, but Heli, or Eli, according to Luke). But the Y genes of an XX male must come from some single male ancestor of Mary, or from several of Mary's male ancestors. The Greek philosopher-theologian Justin Martyr, in his *Dialogue with Trypho,* argued that Mary herself was descended from David.[45] Assume she indeed had some male ancestors who were descended from David—in genetic terms, this means that each such man had a Y chromosome identical to David's, since the Y chromosome does not recombine. Then, either once or several times, insertions could have been made from the Y chromosomes of these men into the X chromosome that was to become the container of Jesus' Y genes. If all (or at least most) of the Y genes of David were present in Jesus, he in fact would be descended from David in the male line in the sense that the term *descent* is applied genetically: the male has the Y genes of the male ancestor. Thus, even though Joseph was not the biological father of

Jesus, Jesus could be a male of Davidic descent, and so it would have been appropriate to give Joseph's line of descent from David.

Note that a genealogy giving the gene insertions could have more than one male supplying a Y gene in the same generation. Or the insertions of a Y gene could occur once in several generations. Thus, to list the complete line of descent correctly, a genealogy would have to list more than one male in the same generation. This could be accomplished by providing two mutually inconsistent lists—but inconsistent only if we do not realize that Jesus was an XX male, born of a Virgin. It is suggestive that Matthew (1:17) insisted—even if he had to miscount[46]—that there are 28 generations from David to Jesus, *exactly* the number of genes that the Y chromosome carries. Luke (3:23–31) lists 42 generations from David to Joseph (Mary's generation), inclusive. Could the insertion of 28 genes require 42 generations? Another interesting possibility comes from the fact that Luke (3:23–38) lists 77 generations altogether.[47] It has recently been discovered that although the human Y chromosome has 28 genes, these genes exist in multiple copies, for a total of 77,[48] in which case it might be that the total number of generations in Matthew refers to the number of distinct genes in the Y chromosome (28), and in Luke, the total number of genes in the Y chromosome (77). Or perhaps this is just meaningless numerology, and Matthew and Luke are just recording different family traditions of descent, both of which are incorrect. After all, Luke is giving 77 generations since Adam and Eve, the first humans, whereas, as mentioned earlier, behaviorally modern humans have existed for at least 60,000 years, which would mean more than 2,000 generations since those first modern humans.

But we would have insufficient hints of how to work the electroweak baryon-annihilation process if we did not have a sample of Jesus' blood and flesh to study. So, following the logic of my Christology to the end, I conclude that such a sample must exist.

The Shroud of Turin

Could it be on the Shroud of Turin? This artifact, pictured in Figure 7.1, a linen cloth with the faint outline of a human figure, is widely believed to be the burial cloth of Jesus.

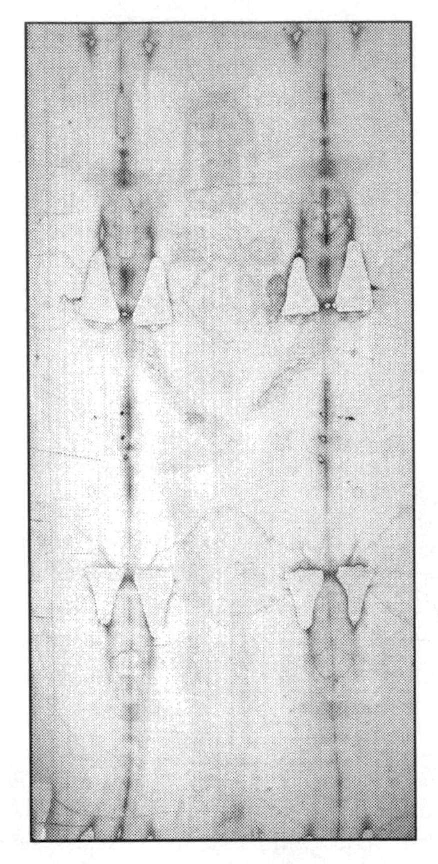

Figure 7.1. Photograph of the body image on the Turin Shroud.

The radiocarbon dating of the Shroud is known to be incorrect, first because bacterial contamination was not taken into account (bacteria add carbon of a later date than the actual Shroud material and thus make it seem younger than it is),[49] and second, because the Shroud samples tested were apparently from a section that had been partially "repaired." The chemist Raymond Rogers has done a careful chemical analysis of linen fibers taken from all areas of the Turin Shroud, and he has shown that it is almost certain that the linen used to obtain the radiocarbon date was medieval in origin.[50] That is, the particular sample taken from the Shroud to obtain its age by radiocarbon dating was not manufactured at the same time as the rest of the Shroud. This fact suggests that the linen from the radiocarbon sample was added at a later date, probably to repair the Shroud. The radiocarbon analysis yielded a date between A.D. 1260 and 1390 completely consistent with Rogers's chemical analysis of the linen fibers from the radiocarbon area.

Linen is made from flax, and flax contains a chemical substance called *lignin.* Over time, the lignin will lose its content of another chemical compound called *vanillin.* Thus, one can obtain an estimate of the age of a linen sample by comparing the relative amount of lignin and vanillin. Rogers detected vanillin in the radiocarbon samples but could not detect any vanillin from the samples from the areas of the Shroud. From this, he inferred that the original Shroud material is between 1,300 and 3,000 years old. Rogers also detected alizarin dye in the radiocarbon sample but no such dye in the original Shroud material. The dye was apparently used to make the color of the repair material match the color of the original Shroud (linen turns yellow over time). This particular dye was introduced in Italy in 1291, so the radiocarbon sample cannot be older than this date. Indeed, the central value of the radiocarbon date, 1325, is about three decades after the date of the first use of the dye in Italy.

Bacteria generated an almost completely transparent bioplastic coating of the linen fibers that make up the Shroud. Leoncio Garza-Valdes, the scientist who discovered this coating, removed it and sent two samples of the decontaminated Shroud linen to two different radiocarbon dating laboratories for a redating.[51] Unfortunately, Garza-Valdes was not an expert in the handling of materials for radiocarbon dating analysis, and he inadvertently added some "dead" carbon to the linen sample. "Dead" carbon is carbon that has not been in the atmosphere for millions of years, and so almost all of its carbon 14 has decayed. The carbon in coal and oil is all dead carbon, for example. Adding such carbon makes the sample appear to be much older than it actually is, so Garza-Valdes concluded that his redating experiment was a complete failure. But the chemist Alan Adler has pointed out that if we use standard chemistry to make a "reasonable" estimate about how much dead carbon from the reagent used by Garza-Valdes would be absorbed by the cellulose of the linen, we can obtain an estimate of the actual radiocarbon date of the decontaminated Shroud linen.[52] The corrected date is A.D. 351, a date consistent with the first century to within the accuracy of the "reasonable" estimate. This unofficial radiocarbon redating does not, of course, establish that the true date of the Shroud is the first century A.D. It does, however, provide justification for believing that the original radiocarbon dates do *not* rule out the Shroud as genuine.

If the radiocarbon date is ignored, there are quite a few reasons for accepting the Shroud as genuine. The best available popular summary of these arguments can be found in the books of Ian Wilson and Barrie Schwortz.[53] Also in this book is the best popular summary of the evidence for bacterial contamination on the Shroud.[54] But what this book does not answer—what must be answered before the Shroud can be accepted as genuine—is why the radiocarbon date is *exactly* what one would expect it to be if the Turin Shroud were actually a fraud.

A very plausible history of the Shroud from A.D. 30 to the present has been constructed, and I shall outline it a bit later in this chapter. However, the first time the Shroud is agreed by all scholars to have existed is 1355, when a French squire, Geoffrey de Charny of Lirey, in the bishopric of Troyes, petitioned the Pope to display it as the unique burial cloth of Jesus. This French city we have encountered before, as the home of Rashi, the greatest of medieval Jewish commentators on the Bible. De Charny never explained how a completely unimportant person such as he managed to obtain possession of the most important relic in Christendom. Even the people living in the Middle Ages, who were often credulous about relics, were suspicious. A few decades after de Charny's death, the bishop of Troyes denounced the Shroud as a fake and said that he knew the name of the forger, who had confessed. So if the bishop and later skeptics were correct, we would expect the linen of which the Shroud is made to date from the time of the forgery. That is, the middle of the fourteenth century. When the radiocarbon date was discovered to be between 1260 and 1390 (95 percent confidence interval), most scientists (including myself until a few years ago) were convinced that the Shroud had been proven a fraud. If bacterial or other contamination had distorted the date, we would expect the measured radiocarbon date to be some random date between A.D. 30 and the present. It would be an extraordinary and very improbable coincidence if the amount of carbon added to the Shroud were exactly the amount needed to give the date that indicated a fraud.

That is, unless the radiocarbon date were itself a miracle in the sense defined in Chapter 5: an improbable event as computed using past-to-present causation, but an event seen to be inevitable using future-to-past causation. I shall argue that this is the case, but for the nonce, let me merely note that, if the Shroud is indeed genuine, we should not be surprised that one more miracle is associated with it.

There are several ways of using current radiocarbon dating technology to date the Shroud correctly. The obvious way is to remove the bioplastic coating. This is difficult to do, and that is why the cleaning protocols of the three laboratories performing the radiocarbon dating failed to remove the coating. Leoncio Garza-Valdes has informed me that it is possible to use sodium hydroxide to dissolve the linen of the Shroud so that only the bioplastic coating remains.[55] If indeed the linen of the Shroud dissolves, leaving *all* of the bioplastic coating behind, then there is a simple procedure to obtain the actual age. First measure the age of the linen plus the contamination (this is the number we already have).[56] Then measure the age of the contamination alone. This should be easy. Most of the mass of the Shroud must be in the form of contamination if the Shroud is genuine. Calculation shows that between 60 and 90 percent of the total mass would have to be in the contamination if a first-century linen shroud were to be mistakenly dated to the fourteenth century. From the two age measurements, one can compute the true age of the linen cloth, even if one does not know when the linen was contaminated. We would in fact expect the Shroud to have been contaminated almost continuously from the first century to the present if the history defended by Ian Wilson and other Shroud scholars who believe in the Shroud's authenticity is correct.

This proposed history is fascinating. In fact, it is nicely and beautifully illustrated in a seventeenth-century icon acquired by Prince Albert (Queen Victoria's husband) in the nineteenth century and now on display in Hampton Court, a royal palace in Great Britain. The original Shroud in which Jesus was buried was taken at some unknown time (but shortly after his Resurrection) to the city of Edessa (the modern city of Urfa, Turkey), where the king of Edessa from A.D. 13 to 50, Abgar V, venerated it and set it on the walls of the city. A later king of Edessa reverted to paganism and began persecuting Christians and destroying holy relics, so the Shroud was hidden in the city walls. The men and women who had hidden the Shroud apparently were killed in the persecution before they were able to reveal where they had put it.

The Shroud lay hidden for centuries, during which time Christianity became the official religion of the Roman Empire and Edessa became a city of the empire. When the city walls were rebuilt in the sixth century, probably after a major flood in 525, the Shroud was rediscovered. In the seventh century, Edessa was conquered by the Muslims, in whose eyes the Shroud was a barely tolerated infidel relic. In 944 a Byzantine army lay siege to Edessa and raised the siege only when the Emir of Edessa agreed to give up the Shroud to the besiegers. They took it to the capital of the Byzantine Empire, Constantinople (the modern city of Istanbul, Turkey). There it remained, known as the Mandylion, a relic with a miraculous image of Jesus' face (most of the time, the only part of the Shroud that was displayed was the face), until it disappeared when the soldiers of the Fourth Crusade sacked Constantinople in 1204. The Shroud thus came under the secret ownership of the Knights Templar. It remained in their hands until 1307, when Philip the Fair, king of France, suppressed the Knights Templar on charges of heresy, part of the evidence being that the knights worshiped in secret the image of a bearded man. If this reference to the "image of a bearded man" was to the image of the face of Jesus on the Shroud, and if the Shroud were genuine, then the worship would have been perfectly orthodox, since they would have been worshiping Jesus.

One of the leading Templars to be burned at the stake in 1314 for heresy was the Templar master of Normandy, Geoffrey de Charny. This is a name we have encountered before: the name of the first man generally agreed to have owned the Shroud. Surviving records are insufficient to establish a family relation between the two men of the same name living a generation and about fifty miles apart. But the image supposedly worshiped by the Templars was never found, and had it been found, it would have been the property of the king of France. Hiding the image from the king would have been treason (if not heresy) and punishable by death. If it was the Shroud that the younger Geoffrey de Charny obtained secretly from the older Geoffrey de Charny, then the reluctance of the former to say where he had obtained what he claimed to be the genuine burial cloth of Jesus is understandable, though it does make it impossible to verify the genuineness of the Shroud by records. Another way will have to be used, and I will describe one.

Leoncio Garza-Valdes owns a mummified Egyptian ibis that has been wrapped in linen. It has been shown that the linen is four to seven centuries younger than the tissue of the ibis, and it is observed that the linen wrap has a bioplastic coating.[57] I call upon Dr. Garza-Valdes to test the bioplastic coating theory by removing the linen from the bioplastic coating and thus measuring the true date of the linen by the indirect method I just described. If the dates agree and are the same as the age of the ibis, the bioplastic theory would be confirmed, and the experiment would provide evidence that might persuade the keepers of the

Shroud to conduct the same procedure on it. Garza-Valdes also has (or had) in his possession samples of the linen wrappings of Manchester mummy number 1770 (a human mummy). The radiocarbon date from the bones of the mummy is 1510 B.C., whereas the date of the linen wrappings is A.D. 255.[58] Garza-Valdes has seen a bioplastic coating on the linen wrapping. The linen of this mummy should also be redated by the indirect procedure.

Garza-Valdes, together with Victor and Nancy Tryon, has carried out a DNA test on blood samples taken from the Shroud.[59] The Tryons conducted a simplified standard sex determination test. He looked for, and found, fragments of the amelogenin-X gene, which is located only on the X chromosome, and the amelogenin-Y gene, which is located only on the Y chromosome. Thus, Garza-Valdes concluded that the blood on the Shroud possessed the full XY chromosome pair. However, his data did *not* have a statement of the actual amount of DNA found of the two types of amelogenin. Thus, the presence of the Y form may just be contamination from the huge number of people who have handled the Turin Shroud over the centuries.

Even if this possibility is ignored, Garza-Valdes's experimental result is also consistent with my hypothesis, that Jesus was an XX male with all of the Y genes present on one of the X chromosomes. In all studied cases, an XX male has only one Y gene, the SRY gene, which is responsible for the testes-determining factor, inserted into an X chromosome. But if my hypothesis is correct, then the way to distinguish Jesus' genes from those of the normal XY male is to look for the SRY gene (or any Y gene if we assume Jesus has all the Y genes) and simultaneously conduct a test for two distinct alleles of the most variant of the X-chromosome genes. Human females will have two alleles of the X-chromosome genes because they have two X chromosomes. The X genes will be the same only for those genes that have only one variant, that is, only one allele.

As I mentioned earlier, there are two other hypotheses for how a Virgin Birth could have occurred. One theory, advanced by the geneticist Sam Berry of the University of London, assumes that Mary was an XXY female (Klinefelter's syndrome). All observed XXY females have undeveloped wombs, but under this hypothesis, Mary was on the extreme end of a Gaussian distribution for XXY females, so that her womb was normal. Jesus grew from a cell in which one of Mary's X chromosomes was deleted.[60] Alternatively, Garza-Valdes's own hypothesis, the third hypothesis, is as follows. A tumor in the form of an undeveloped male embryo was in Mary's womb from her birth. As Garza-Valdes points out, such embryos (at least in the XX variety) have been reported in the medical literature, and he himself had a patient with this abnormality. The embryo in Mary's case would have fertilized one of her eggs, resulting in the Virgin Birth of Jesus.[61]

The Virgin Birth under Garza-Valdes's theory would be a virgin birth in the sense that Mary would have had a child without having sexual intercourse with a man, but it would genetically be a brother-sister mating: the embryo that fertilized Mary was really her undeveloped brother. An incest mating has a genetic signature: since the mother and father are close relatives, they would have many of the same alleles for the same gene, and thus the child of an incest mating shows much less genetic variability. One religious problem with Garza-Valdes's theory is that his form of virgin birth cannot be distinguished from an actual mating of Mary with her brother (if such existed). The charge of brother-sister incest was apparently leveled at Mary in Alexandria early in the Christian era.[62] If a DNA analysis yielded the result indicated by Garza-Valdes's theory, the charge of incest would once again be raised. For this reason I personally find (and I suspect orthodox Christians in general would find) the Garza-Valdes theory morally repulsive. And, of course, it is inconsistent with the assertion that Jesus was conceived by the power of the Holy Spirit. However, we must keep in mind that finding a theory repulsive does not mean that it is not true. Truth must be decided by experiment. The XXY hypothesis would yield a male who genetically looks normal, is morally acceptable, and is consistent with the Holy Spirit assertion.

So these three hypotheses for the Virgin Birth can be distinguished from one other by the appropriate DNA test. The required test will admittedly be difficult to carry out, since, according to Garza-Valdes, bacteria have replaced 95 percent of the blood on the Shroud.[63] But the experiment can be carried out. The test for an XX male would be easier, since it would involve carrying out the two standard tests for maleness: test for the SRY gene (or possibly, any Y gene) and simultaneously test for two alleles of several X-chromosome genes.

Mark Guscin provides strong evidence that the Sudarium of Oviedo, Spain, is the cloth described in John 20:7 as being wrapped around Jesus' head.[64] A DNA test for the three distinct ways of having a male born of a virgin should therefore also be carried out on the Oviedo

Cloth. Photographs of the blood on the Turin Shroud and on the Oviedo Cloth are pictured in Figures 7.2 and 7.3, respectively.

DNA Test of the Blood on the Turin Shroud and the Oviedo Cloth

In January 1995 a group of Italian researchers, led by Professor Marcello Canale of the Institute of Legal Medicine in Genoa, conducted a DNA analysis of the blood on the Shroud. This group included several workers who had invented the standard DNA test for gender. And their experiment was much more complete than the one conducted by Leoncio Garza-Valdes, since it included detailed information about the amount of DNA obtained. Contamination can thus be quantified.

This group simultaneously tested the blood on the Oviedo Cloth.

I had great difficulty acquiring a copy of their article[65]—Tulane University Library was unable to obtain a copy, and this failure is very unusual—but the Turin Shroud researcher Barrie Schwortz put me in contact with Ian Wilson, who e-mailed me a copy. I was surprised at my difficulty. Normally, the results of a DNA test of the blood on such a famous object would be published in English in a major scientific journal.

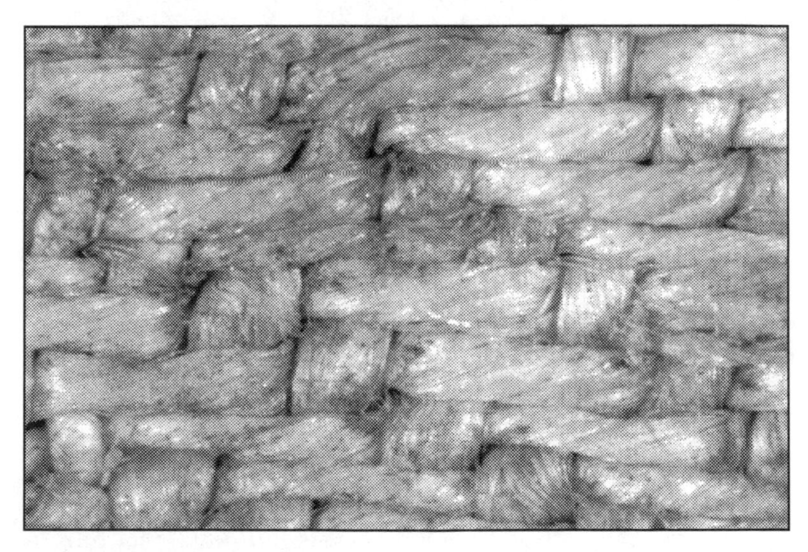

Figure 7.2. Close-up of the bloodstains on the Turin Shroud.

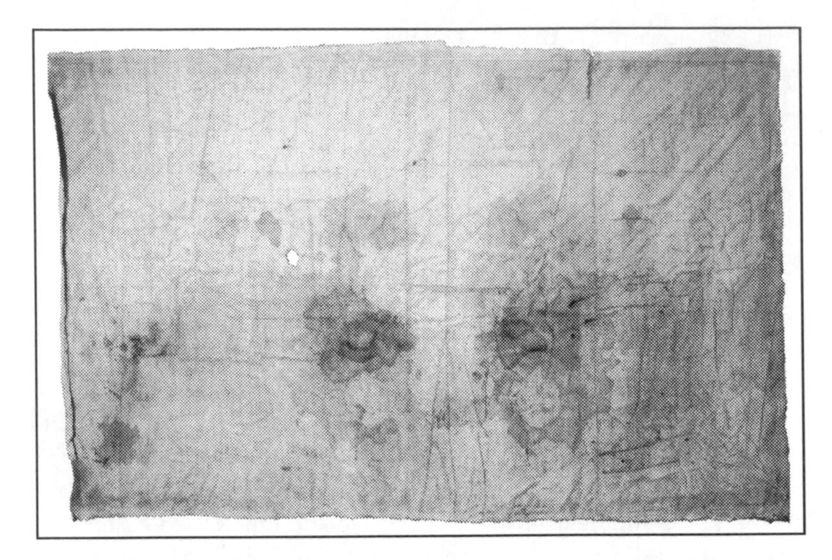

Figure 7.3. Photograph of the bloodstains on the Oviedo Cloth.

For example, the results of the DNA test establishing that Thomas Jefferson fathered children on his slave Sally Hemings were published in *Nature,* the world's leading science journal. English has become the standard language of science, so all scientists today, whatever their nationality, would almost certainly publish important findings in English, if for no other reason than to be sure scientists everywhere would be able to read the article.

Not so the results of this DNA test. The results were published, in Italian, in the very obscure Italian journal devoted to the study of the Turin Shroud. Furthermore, only the raw data were published. That is, the Genoa team published black-and-white Xerox copies of the computer output of the DNA analyzer. This is never, never done. Always, the data are presented in a neat table or figure, and they are accompanied by a discussion of their significance. The Genoa team made no effort to interpret their data. The combination of these four facts—obscure journal, non-English language, raw data only, and no attempt to interpret the data—almost certainly means that the researchers thought the data to be worthless, incapable of interpretation.

But I was able to interpret the data at once. *They are the expected signature of the DNA of a male born in a Virgin Birth!* The data are presented in standard tabular form in Tables 7.1 and 7.2.

Table 7.1

The DNA on the Turin Shroud

Min (peak label)	Size (in base pairs)	Peak Height	Peak Area	Scan Number
77	82.53	21	128	771
85	96.26	41	511	853
90	107.28	27	286	903
105	130.33	28	279	1052
105	131.16	35	255	1057
123	161.61	26	351	1231
129	174.14	20	113	1298
130	174.68	20	110	1301
152	212.37	52	418	1527
153	213.04	48	340	1531
155	216.40	98	1258	1551
157	220.63	55	720	1576
159	225.03	59	686	1599
162	230.04	33	319	1625
165	234.56	57	682	1651

Table 7.2

The DNA on the Oviedo Cloth

Min (peak label)	Size (in base pairs)	Peak Height	Peak Area	Scan Number
75	74.89	65	313	759
93	105.27	55	298	935
127	161.42	65	707	1273
162	221.24	53	472	1627

The standard DNA test for sex is the amelogenin test I mentioned earlier.[66] The Italians performed this test, which gave 106 base pairs for the X form of amelogenin and 112 base pairs for the Y form. There is a phenomenon called *sputtering,* which can cause the actual value obtained to differ by 1 base pair from the expected value.

The Turin Shroud data show 107 (106 + 1) but no trace of a 112 base pair gene. The Oviedo Cloth data show 105 (106 − 1) but no trace of a 112 base pair. The X chromosome is present, but there is no evidence of a Y chromosome. This is the expected signature of the simplest virgin birth, the XX male generated by an SRY inserted into an X chromosome. It is *not* what would be expected of a standard male.

This is the DNA signature of the *simplest* way of generating an XX male. The more complicated proposal for an XX male can be rejected, together with the alternative proposals made by Leoncio Garza-Valdes and Sam Berry.

Other explanations are possible. The DNA analyzed could be entirely contamination from people who later touched the Shroud and the Cloth. But we have witnesses that men touched the two samples also, and it seems incredible that no trace of male contamination would be seen. Also, the Italian researchers were aware of the possibility of contamination, and they took precautions to make sure that they analyzed the DNA of only the blood samples. Another possibility is that the Turin Shroud and Oviedo Cloth are fakes and that the fakers used real blood from males they knew were born of virgins. This possibility, in my opinion, has zero probability.

Nevertheless, there is evidence that can be interpreted as contamination in the DNA from locations other than the X and Y chromosomes. In addition to the sex determination genes, the Italian researchers used what is now called a "first generation multiplex." They looked for alleles of the genes TH01, VWA, FES/FPS, and F13A1. The first two genes have twenty and the second twenty-nine known alleles in the human population.[67] The last two genes are no longer used in standard DNA tests. If the DNA came from only one individual, we would expect to see at most eight different alleles from these four genes, since each person has two copies of each gene. There are fourteen distinct alleles seen in the Turin Shroud data (once the AMEL gene is subtracted), so the most obvious interpretation is that there are six more alleles present than could be from a single individual.

The Italian researchers made a passing remark that this finding indicated contamination. This would be the standard interpretation. However, by hypothesis, this sample is DNA from an XX male, and the signature of such a male is the insertion of DNA into a location where it is not normally found. It is therefore possible that additional insertions could have been made at other locations, resulting in more than two alleles present. If there were contamination from many individuals, we would expect to see more than six additional alleles. There is also the possibility that some DNA strands in the sample were degraded, resulting in the appearance of several alleles where only one was present in the original sample. Unfortunately, the published data do not allow anyone to determine which alleles are present. The genes are coded with colored dyes, so that they appear as different colors in the raw data. But the article in which the data were published was in black and white, and thus this crucial color information did not appear. The essential point is that contamination of the sample with genes other than the AMEL gene would not affect the conclusion that there is no evidence of a Y chromosome present.

Standard DNA tests now (remember that the Italians did their analysis in 1995) include tests for base pair sequences on the SRY gene complex. This test should be carried out before we say that the blood on the Turin Shroud and the Oviedo Cloth is that of a male born of a virgin. We should also keep in mind that there is one reported case of an XX male without the SRY gene. This is possible, since genes on chromosomes outside the X and the Y genetically determine maleness. The genes of the X and Y merely tell these genes to turn on or off.

The Italian data were not what the researchers expected. In my experience as a scientist, a correct result obtained by an experimenter who did not expect that result can be depended on. The DNA data thus support the virgin birth hypothesis. The DNA data supporting a virgin birth also support the hypothesis that both the Turin Shroud and Oviedo Cloth are genuine.

The Immaculate Conception and the Fall

In principle, it might be possible to show, by DNA analysis of the blood on the Shroud (or the Sudarium, another name for the Oviedo Cloth), not only that Jesus was born of a virgin but that he and his mother were without Original Sin. That is, it might be possible to confirm the Immaculate Conception (Catholic Church dogma since 1854). In the Christian tradition, Original Sin is inherited from our ultimate ancestors. If Original Sin actually exists, then it must in some way be coded in our genetic material, that is, in our DNA. Also according to the Christian tradition, Original Sin originated in the Fall, an act of some kind by our ultimate ancestors. Before the Fall there was no sin. Almost all scientists consider the Fall a fairy tale. I want to argue the contrary. I shall claim that there was a time in Earth's history when no sin or evil existed, that sin came into the Earth's biosphere at a definite time in the past, and that not only we humans but all metazoans are infected by it. A tendency to commit evil acts is indeed in our DNA, and hence it is inherited. But this tendency might not be present in all humans' DNA. A man and a woman *might* not have had the sin behavior genes.

We first have to have a clear conception of "sin" or "evil." In modern English, *sin* refers to an offense against sexual morality. An example would be concupiscence, or excessive sexual desire. But of course *evil* is a more general phenomenon. In the Judeo-Christian tradition, sex itself is good because it was created by God, as asserted in Genesis 1:28 and in Genesis 1:27 (when read in conjunction with Genesis 1:31). The natural goodness of sex is also asserted by Paul in 1 Corinthians 7:38. The claim that sexual relations between man and wife are intrinsically bad is the Gnostic heresy. Instead, Paul in Romans 7:7 argues that covetousness is the foundation of all evil.

But covetousness is not desire per se. It is perfectly correct to desire the happiness of other people, for example. It is also perfectly legitimate to desire to increase one's knowledge. In particular, Eve's desire for knowledge (Genesis 3:6) was not evil. Only the "knowledge" of evil as distinct from "knowledge" of good (Genesis 2:17) is evil. (*Knowledge* is the sense of information coded in the genes.) As the Tenth Commandment (Exodus 20:17) tells us, desire becomes evil only when one desires someone else's property. Then desire becomes "covetousness." The last six Commandments, the Commandments concerning interhuman relations, can be summarized by saying, "Don't take, don't even think of taking, something that belongs to someone else." Don't take away the respect due to your parents, don't take someone's life, don't take someone's mate, don't take someone's property, don't take someone's reputation. Don't even *think* of taking these things. The first four Commandments tell us to love and respect God. Jesus agreed (Matthew

19:19 and Luke 10:27) that the essence of evil is failing to love God and not respecting others' rights (not loving others as oneself).

I claim that we can summarize these definitions of evil in the following way: All evil acts can ultimately be reduced to a violation of one and only one ethical rule: *Thou shalt not impose thy theories on other living beings by force.* Thus, moral evil is a certain type of interaction between two or more living organisms. Natural evils are of two types: pain (both physical and mental) and death. So a world without evil is a world in which there is no death, no pain, and no force applied by one organism on another.

This statement actually describes the world of the one-celled organisms 2 billion years ago. As emphasized by Lynn Margulis and Dorion Sagan, before the evolution of metazoans, there were no distinct species.[68] All one-celled organisms can exchange genetic information with one another. Furthermore, for each type of microorganism, there were many clones. Since no information was coded except in the genes, these apparently distinct organisms were really "backup" copies of one another. The "individual" was really not a single cell but rather the collection of all clones of that cell. As long as a single cell of the clone existed, the individual existed. Certain lines of cyanobacteria have not changed for over 3.5 billion years. This individual has been alive almost as long as the Earth itself has existed. One-celled organisms have no nervous systems, so they cannot feel pain, either physical or mental. It is not possible to apply force to an organism that cannot feel pain and cannot die.

Applying force—evil—became possible with the evolution of the metazoans. Information was now coded in relationships between the cells, as in the nervous systems of chordates. This information was unique to the individual, not just the clone. It could be destroyed. Death and pain entered the world and, with them, the possibility for moral evil. A metazoan *could* impose its will on other organisms. It could impose its theories on other organisms. One way would be to eat these other organisms. The information coded in the eaten organism would disappear and be replaced by information coded in the eater. This is a simple example of theory imposition. We humans are more memes—ideas, complexes coded in nervous systems—than genes, so we are more familiar with forcible theory replacement of meme than of gene. But both gene and meme replacement are examples of theory replacement.

By the time of the Cambrian Explosion, if not earlier, carnivores had appeared on Earth. Evil had appeared in the world. Genes now coded for behavior that guided the use of biological weapons of the carnivores. The desire to do evil was now hereditary.

We humans ourselves show a marked tendency to want to impose our will on other organisms, both our own species and others. I would suggest that this tendency is genetic, as it certainly is in other meat-eating animals. But we are omnivores: the human per capita murder rate—the killing of members of one's own species—is less than that observed in pure carnivores, such as lions and wolves. So we are not the most violent, the most evil, of animals. We also have no reason to think that this violent tendency is absolutely essential to our survival. Pacifists such as Mahatma Gandhi have insisted that it is not, but even Gandhi occasionally showed a wish to impose his will on others. The genes that generate this tendency are probably universal in the human species.

But if these genes are not essential to human survival, we can imagine that they could be absent from some perhaps unique individuals. The Judeo-Christian tradition holds that the female began the Fall. Since it is absurd to think that the Fall began with a female because females are less able to resist temptation than males (indeed, Genesis 2 does not suggest this), it is more reasonable to interpret the tradition to mean that the gene essential to the evil tendency is on the X chromosome. It may be relevant that the damage of a certain X chromosome gene is known to be responsible for violence in males. But too little is now known about behavioral genetics to say definitely where the evil tendency genes are located. Or, I admit, even if they exist. If they do exist, though, the Christian tradition would claim that these genes would be absent from Jesus' genome.

Since Jesus and Mary would share the same genome on my XX male theory, if the genes are absent from Jesus' genome, they would be absent from Mary's. Jesus would indeed have been conceived immaculately. A DNA search in the Shroud for the X-chromosome gene just mentioned would be a first step. If this gene were indeed involved in our tendency to commit evil, we would expect to see this gene modified from the human norm in the Shroud DNA. In fact, if the evil gene is connected to bone generation, the amelogenin gene, which codes for the generation of teeth, might be entirely absent from Jesus' genome both in its X form and in its Y form. If so, this gene would be absent from the DNA on the Shroud of Turin if this artifact is genuine. If the Christian tradition is correct that the Fall affected the entire animal kingdom, we would expect to see a similar evil gene complex present in all animals, presumably in the chromosome coding for sex differentiation. In mammals this is the X chromosome (if both males and females are to be subject to the gene), but in birds it would be the Z chromosome. In reptiles, it would be present in both sex chromosomes, since in reptiles sex is determined by the temperature experienced by a given egg rather than the genes, as described earlier.

If the evil tendency genes were implanted in the genome by eating something, as Genesis 2:16 claims, then the Fall would have occurred near the start of metazoan evolution. As Margulis has shown, ingestion is a common way for one-celled organisms to obtain new genes, and we would expect this capacity to persist only for very early metazoans.[69] If the Fall occurred at the time of the Cambrian Explosion, I would conjecture that the gene is associated with the formation of bony substances, since such materials were used to form weapons in creatures that lived at that time. I find it very intriguing that animals which have been domesticated have more delicate bones than do the lineages from which they originated.[70] This indeed suggests that bony structures are in some way associated with the ability to fight rather than to cooperate.

If a gene for evil exists, there would have to be a reason for it being universal in the metazoan world. If this gene appeared half a billion years ago, why was it not deleted in some lineages? It would not be deleted if it also coded for some characteristic that is completely essential for metazoan life, so that eliminating the gene for evil would also delete a characteristic for metazoan life. This is another reason for suspecting that the gene might be associated with the formation of bone. Reconstruction of the gene complex for bone formation would be too improbable or difficult. We would thus expect the reconstruction of the complex to have occurred only once, that is, in Mary and Jesus.

If an evil gene complex is indeed ubiquitous in the metazoan biosphere, a truly existing Satan would be possible: he is an evil program coded in the biosphere. He is coded sometimes as a gene complex, sometimes as a meme complex, and sometimes as both. (In a meme the information is coded in a nervous system rather than in the genetic system. A meme is thus an idea complex that is passed from one individual to another by means such as vocal communication.) The evil program thus tempts us to impose our theories on others. Jesus, of course, could be tempted only by a meme: he knew about his ability to impose his will on others by using his dematerialization power, as I shall discuss in the next chapter. Using this power, he could easily have conquered the world. He was tempted to do so, but this would have meant worshiping Satan—accepting evil by carrying it out. This picture of Satan resembles that put forward by C. S. Lewis in his Space Trilogy, and it raises the fascinating possibility that there may be other biospheres, possibly with intelligent life, that never underwent the Fall. The laws of physics require that there are other biospheres with the capability of dematerialization. Perhaps these biospheres never fell. As pointed out by Lewis, the Christian tradition implies that our biosphere is unique in having fallen: Jesus was incarnated only once in the universe.

If Jesus' conception was Immaculate, then it is very appropriate to call Mary *Theotokos,* the Greek word being most accurately translated as "the one who gave birth to the one who is God."[71] An Immaculate Mary would be both completely human and more than human: she would be missing the genetic flaws that induce us to do evil. Since she is not God, she is not entitled to the worship that God (the Father, the Son, or the Holy Spirit) is entitled to. The Catholic Church uses the word *latria* ("adoration") for this form of worship. A mere saint (a normal human with Original Sin but who has more or less managed to overcome this inducement to commit evil) is entitled to reverence, or *dulia.* Mary, being more than a saint but infinitely less than God, is entitled to *hyperdulia,* the prefix *hyper* meaning "reverence to the highest degree."[72] If, as I shall suggest in Chapter 8, Mary's genetic constitution allowed her to be assumed into heaven, then calling her *Theotokos* is doubly appropriate.

In the preceding discussion, I have assumed a normative principle—Thou shalt not impose your ideas on others—to simplify the analysis, but actually all moral rules, including this one, can be derived from facts alone, which ultimately means from physics alone. Coase's Theorem, which won Ronald Coase the Nobel Prize, asserts as much.[73]

If Jesus was indeed an XX male, as the DNA on the Turin Shroud and Oviedo Cloth strongly suggest, then presumably he would share a common characteristic of XX males: not being fertile with normal human females. But we would expect this infertility with normal humans solely on the basis of him (and possibly his mother) being without Original Sin. The genetic basis of Original Sin is so fundamental in normal human beings that he and his mother should be regarded as a new species: humans without Original Sin or, as the story of the Garden of Eden makes clear, what we humans were intended to be, not what we actually are. As such, Jesus would indeed be "a lamb without blemish, and without spot" (1 Peter 1:19).

From the biological point of view, we would have in Jesus a speciation event, the appearance of a new species in a single generation. Mary's parents were normal humans, whereas Jesus and Mary were the new Adam and the new Eve. No Darwinian slow evolutionary change here, but instantaneous speciation. This would explain the fact that the Gospels make no mention of Jesus ever taking a wife, or showing any interest in women as sex objects. So the DNA on the Turin Shroud and the Oviedo Cloth provide an experimental refutation of the claim, made popular in the novel *The Da Vinci Code,* that Jesus married and had children by Mary Magdalene. Jesus' DNA is simply too different for this to be possible.

VIII

The Resurrection of Jesus

> Now if Christ be preached that he rose from the dead, how say some among you that there is no resurrection of the dead? But if there be no resurrection of the dead, then is Christ not risen. And if Christ be not risen, then is our preaching vain, and your faith is also vain. Yea, and we are found false witnesses of God; because we have testified of God that He raised up Christ: whom He raised not up, if so be that the dead rise not. For if the dead rise not, then is not Christ raised: And if Christ be not raised, your faith is vain; ye are yet in your sins. Then they also which are fallen asleep in Christ are perished. If in this life only we have hope in Christ, we are of all men most miserable.
>
> 1 CORINTHIANS 15:12–19 (KJV)

The Case Against the Resurrection

Before discussing a physical mechanism for Jesus' Resurrection, I want to present the skeptical case *against* the Resurrection. The skeptics have not made a strong case against the Resurrection because they have tacitly assumed that their readers will "know" that Jesus' Resurrection is impossible as physics. If a reader already believes that Jesus' Resurrection is impossible, it is not necessary to offer a convincing explanation of what really happened in A.D. 30. But I shall show that Jesus' Resurrection *is* possible as physics. If this is so, then the alternative, that Jesus did not rise from the dead, must be defended on its own terms, and an explanation of why the early Christians acted as if they truly believed Jesus did rise has to be provided. Christians reading this skeptical case should not be dismayed. The truth can stand any criticism.

Modern Christians have provided convincing evidence of several facts.[1] The first, already mentioned, is the overwhelming evidence that Jesus' disciples really and truly believed they had seen the Risen Jesus, talked to Him, and touched Him. All historical evidence indicates that the disciples went cheerfully to their deaths, often horrible deaths, because they believed they had seen the Risen Jesus. Second, the evidence is strong that Jesus' tomb was empty. If it were not, then it would have been a simple matter to open the tomb and present the dead body of Jesus, conclusively refuting the claim that he had risen. Third, the empty tomb claim was unlike empty tomb claims in ancient times, and descriptions of the Risen Jesus in the Gospels are unlike the resurrection claims of the gods in ancient religions.

However, it must be kept in mind that the disciples did not see the Risen Jesus at the instant they died in the Roman circus or on the cross. Rather, they went to their deaths with the *memory* of having seen the Risen Jesus. This crucial point allows the strongest skeptical argument to be made.

Most people, including all biblical scholars and most skeptics, assume that human memory is similar to a camera or a camcorder. The human memory, it is assumed, makes a reliable recording of what the person actually saw. Memory scientists know this to be false. Human memory is incredibly malleable, even on important matters. Human memories of what happened and videotapes of what happened can be completely different. More significant, there are social mechanisms that can cause several people's false memories of an event to agree with one another yet disagree with what a videotape shows actually happened. The skeptics would use these facts to argue that Jesus' Resurrection was a false memory of seeing the Risen Jesus, with the social mechanism acting to ensure the false memory is consistent among the disciples. Such an explanation of the general belief among the disciples of having seen the Risen Jesus is far more plausible than alternatives, for example, that the disciples suffered a mass hallucination. Christian apologists have correctly pointed out that there is no evidence of such a phenomenon.

In contrast, there is a case of what courts have decided is a false memory, of several witnesses agreeing that they saw a man whom they believed to be dead, living again. This is the case of John Demjanjuk, claimed to have been the Treblinka death camp guard Ivan the Terrible. There was an uprising at Treblinka in 1943, during which several Jewish prisoners escaped, carrying with them the memory that Ivan the Terrible was beaten to death with a shovel. Several of these prisoners testified right after the war that Ivan the Terrible was killed. And so it was believed until 1980, when the Soviet government produced a document indicating a "John Demjanjuk" at the Trawniki, Poland, training camp where the SS taught prison guards for the Sobibor and Treblinka extermination centers. In 1952 John Demjanjuk had emigrated to the United States, where he became a naturalized citizen. In 1981 the U.S. Department of Justice stripped Demjanjuk of his American citizenship on the grounds that he was a war criminal. Demjanjuk denied having ever been in either Sobibor or Treblinka, as a guard or a prisoner. He appealed the denaturalization, but his appeals were denied. In 1986 he was extradited to Israel to be tried as a war criminal.

Pictures of Demjanjuk taken at the time of his immigration to the United States were sent to Israel, where he was identified as Ivan the Terrible by nine of the seventeen eyewitnesses who survived Treblinka and had known Ivan the Terrible. What is important is how the nine eyewitnesses recognized him. It is important to keep in mind that these eyewitnesses really and truly believed that Demjanjuk was Ivan. They later testified to this belief in Israeli court, when Demjanjuk was on trial for his life. If convicted, he would be sentenced to death, and the eyewitnesses knew this. They did not want to send an innocent man to his death. After all, their innocent relatives had been killed by the Nazis, and the eyewitnesses did not want innocent blood on their hands.

The story of the eyewitness identification process can be found in an important book on false memory, *Witness for the Defense,* by Elizabeth Loftus and Katherine Ketcham. On May 9, 1976, the Treblinka survivor Eugen Turowski did not recognize a picture of Demjanjuk in a montage of photos. At 1:00 P.M. on the same day, another Treblinka survivor, Abraham Goldfarb, said the photo of Demjanjuk seemed "familiar," but he did not mention the name of Ivan. At 2:30, Goldfarb made a second statement, wherein he identified Demjanjuk as Ivan the Terrible. This must have been a shock to Goldfarb, because right after the war he had written that Ivan was killed in the uprising. Then Turowski was reinterviewed by the Israeli investigators and asked if he remembered a Ukrainian by the name of Ivan Demjanjuk. Turowski replied, "When asked if I knew a Ukrainian by the name of Demjanjuk, Ivan, I declare as follows. I know the name Demjanjuk and even better, the name of Ivan. To me, he was Ivan. This Ukrainian I can well remember. I knew him personally, because at times he came to the shop to have things repaired." On being again shown the montage of photographs, Turowski pointed to the photo of Demjanjuk and said, "This is Ivan. Him I recognize immediately and with full assurance."[2]

How is it possible that Turowski could recognize Ivan "immediately and with full assurance" if earlier he had not recognized him at all? Loftus points out that Goldfarb and Turowski knew each other, testified within hours of each other, and probably talked with each other about Ivan the Terrible being alive. The next day another witness, Elijahu Rosenberg, identified Demjanjuk as Ivan. However, after the war, in 1947, Rosenberg, like Goldfarb, testified that Ivan had been killed in the 1943 revolt. The suggestion is that Rosenberg's identification, like Turowski's, came after Goldfarb's and possibly after Rosenberg had talked with the other two. The next positive identifications of Demjanjuk as Ivan the Terrible occurred in September and October 1976, shortly after the August reunion of Treblinka survivors, where the identification of Ivan by Goldfarb, Turowski, and Rosenberg can be assumed to have been a topic of discussion. They were expecting to see the risen Ivan the Terrible, and they did.

One dominant personality, Goldfarb, generated eight other eyewitness accounts that agreed with his. And all nine accounts were honest. All nine men really believed that Demjanjuk was Ivan. The court found Demjanjuk to be Ivan and sentenced him to death. The Israeli Supreme Court found the evidence insufficient and overturned the conviction, to the distress of the eyewitnesses. Demjanjuk is back in the United States, with his citizenship restored but still fighting Justice Department claims that he was a guard at Sobibor. The Treblinka charge has been dropped. No court accepts Demjanjuk as Ivan.

The Risen Jesus could have been a similar phenomenon. The tomb was empty, and this empty tomb required an explanation. A single powerful personality, St. Peter perhaps, could have believed he saw the Risen Jesus and persuaded the others, just as Abraham Goldfarb convinced eight others.

A vast number of similar examples of false memory have convinced memory scientists that human testimony cannot be trusted and that hard physical evidence must be provided in a trial.[3] If Jesus rose from the dead, then this fact is vastly more important than the guilt or innocence of anyone in a trial. Belief in Jesus' Resurrection should be based on physical evidence.

The image on the Turin Shroud may provide the necessary physical evidence that Jesus rose from the dead in a manner that resembles our own resurrection in the computers of the far future, and establishes Him as Christ. I shall now explain how the dematerialization mechanism electroweak quantum tunneling, which I outlined in Chapter 2, would naturally generate the image on the Shroud. To do this, I shall first outline in more detail how this mechanism works. Then I shall show how it explains *all* the observations of the Shroud image made to date. These are: first, the three-dimensional images obtained by a VP-8 analyzer;[4] second, the fact that the image is located on the uppermost fibers of the Shroud linen; and third, the fact that the image appears to be generated by conjugated carbonyl bonds.[5] I shall finally describe how my hypothesis can be tested experimentally. In other words, I shall describe how we can establish experimentally that Jesus rose from the dead in a manner that confirms His Sonship.

A Scientific Explanation and the Resurrection

Wolfhart Pannenberg wrote in his 1966 book *Jesus: God and Man* that "the possibility of the historicity of Jesus' Resurrection has been opposed on the grounds that the resurrection of a dead person even in the sense of the resurrection to imperishable life would be an event that violates the laws of nature. . . . [But] only a part of the laws of nature are ever known."[6] Pannenberg showed great prescience: the law of physics responsible for Jesus' Resurrection was first discovered in 1976 by Gerardus 't Hooft, who was awarded the Nobel Prize in physics in 1999.[7] This new law was not fully understood until the 1980s. The new law is a consequence of the Standard Model of particle physics. For those not familiar with the Standard Model, the best introduction is Gordon Kane's 1993 book *Modern Elementary Particle Physics.* The new law is described in detail in most books on the Standard Model of particle physics.[8] Steven Weinberg gives a particularly elegant derivation of the new law from the Atiyah-Singer index theorem in his treatise on quantum field theory.[9]

It is this mechanism of baryon annihilation via electroweak tunneling that could have been used to accomplish *all* of the miracles described in the Gospels, in particular the Resurrection. I pointed out in Chapter 3 of this book and in somewhat more detail in my earlier book, *The*

Physics of Immortality, that Jesus' Resurrection body, as described in the Gospels, has all the essential properties of the computer emulation resurrection bodies we all will have in the far future. The property most difficult to duplicate at the lowest level of implementation is the sudden dematerialization (vanishing from the appearance of His disciples) and rematerialization (suddenly appearing inside a locked room).

Dematerialization can be accomplished by electroweak quantum tunneling, which violates baryon number and lepton number conservation. The key reaction would be proton plus electron goes to neutrino plus antineutrino. This would convert all the matter in Jesus' body into neutrinos, which interact so weakly with matter that a person in a room with Jesus would see Jesus vanish. (If the matter of a human body were converted into photons rather than neutrinos, this would be equivalent to the detonation of a 1,000-megaton H-bomb, assuming Jesus weighed 178 pounds.[10] The people of Judea would notice this, though the disciples would not, since they would be vaporized.) Reversing the process could carry out materialization apparently out of nothing. The Resurrection is then merely an example of the first dematerialization of Jesus' dead body, followed by the materialization of a living body. The Resurrection, in other words, is a process profoundly different from the mere resuscitation of a corpse.

This dematerialization and materialization process is enormously unlikely to occur if the probability is calculated in the usual past-to-future causation language. Its probability is calculated as follows. We must start with the probability that the tunneling process occurs in a time interval sufficiently short that disciples would see it as "instantaneous" (one-hundredth of a second). This probability is 10^{-100}. We must then raise this enormously small number to a power equal to the number of atoms in a human body, something like 10^{29}. It is a virtual certainty that no one will ever observe dematerialization of even a single atom via this process. But this calculated probability assumes that the dematerialization is merely a random process, unrelated to the universe at large. If, on the contrary, the universe *requires* the dematerialization-materialization of Jesus to have occurred in order for the universe to evolve into the Omega Point, then the probability is not the gigantically small number I just computed. Instead, the probability is 1. That is, the event is certain to occur.

All eight of the "nature" miracles of Jesus could have been accomplished via the electroweak quantum tunneling mechanism. For example, walking on water could be accomplished by directing a neutrino beam created just below Jesus' feet downward. If we ourselves knew how to do this, we would have the perfect rocket! A simple calculation shows how to support a mass against the force of gravity using a directed neutrino beam.

If Jesus had a mass of 178 pounds, or about 80.8 kilograms (I shall justify this mass shortly), then the force that must be exerted to support his weight against the force of gravity is $F = Mg = (80.8 \text{ kg})(9.80 \text{ m/sec}^2) = 792$ newtons. But the force is the momentum p carried away by the neutrinos per unit time, and for nearly massless particles, such as neutrinos, the momentum equals the energy divided by the speed of light. But if the energy of the neutrinos comes from the annihilation of matter, then this energy equals the mass of the matter annihilated times the speed of light squared ($E = mc^2$). Thus $p/t = (E/c)/t = (mc^2/c)/t = mc/t = Mg$. Thus, the amount of mass that must be annihilated per second, or m/t, must equal $Mg/c = (792 \text{ Newtons})/(3.00 \times 10^8 \text{ m/sec}) =$ 2.64 milligrams per second. Thus, if the field responsible for converting matter into neutrinos extends a short distance into the water below Jesus' feet, and if this field is capable of directing all the neutrinos downward, Jesus would walk on water. Or ascend into the clouds after His Resurrection.

Creation of loaves and fishes is just materialization, as is converting water into wine. (However, materialization is so much more difficult than the reverse, for reasons that I shall discuss at length shortly, that I think it more likely that the loaves and fishes event is merely the result of food sharing by Jesus' audience.) The Transfiguration—emission of light from Jesus' body—could have been accomplished by the emission of photons rather than neutrinos. It is very suggestive that all of Jesus' physical miracles can be accomplished the same way. The apparent exception is the Virgin Birth, but I shall argue that this form of birth is necessary to make it easier for Jesus to dematerialize matter, and hence resurrect Himself.

The new law is sometimes called *electroweak baryogenesis,* or, since at high energy the process of baryogenesis is dominated by field configurations termed *sphalerons,* it is also called *sphaleron baryogenesis.* The word *baryogenesis* refers to the generation of baryons (particles such as protons and neurons) and leptons (particles such as electrons and neutrinos) out of energy states. But in physics, a process can be reversed, and I shall be assuming that the process worked in reverse—baryons and leptons annihilated—in the case of Jesus' Resurrection.

Electroweak baryogenesis conserves $B - L$, where B is the baryon number and L is the lepton number. Thus, a hydrogen atom can be converted into a neutrino-antineutrino pair via the sphaleron reaction $p + e \rightarrow NZ \rightarrow N(\nu + \bar{\nu})$, where p is a proton, e is an electron, Z is the neutral $\bar{\nu}$ intermediate vector boson, ν is a neutrino, and $\bar{\nu}$ is an antineutrino. N is the number of neutrino-antineutrino pairs produced in each annihilation of a (p + e) pair. We shall see shortly that the observations can fix N and show that N is greater than 1. But for simplicity of explanation, I shall assume for the moment that $N = 1$.

This reaction conserves $B - L$. To see this, proceed as follows. The proton has +1 unit of baryon number (and 0 for its lepton number) while the electron has +1 unit of lepton number (and 0 for its baryon number). Thus, initially we have $B - L = 1 - 1 = 0$. The neutrino has +1 unit of lepton number, as does the electron (and 0 for its baryon number), and the antineutrino has -1 units of lepton number (and 0 for its baryon number). Thus, in the final state we have $B - L = 1 + (-1) = 0$. Hence, before and after the reaction, $B - L = 0$. The Z boson has 0 baryon and lepton number. This Z particle is called a *virtual particle.* We shall see soon why the Z has to be present. By conservation of energy, the mass-energy in the hydrogen atom—approximately 1 GeV, or 1 billion electron volts of energy—would be carried away by the neutrinos. If we ignore any other reactions occurring simultaneously, the energy would be equally divided among all the neutrinos. Each would have ½ GeV of energy.

Atoms more complicated that the most common isotope of hydrogen have neutrons. These neutrons can be annihilated into neutrinos and antineutrinos by the following reaction: $n \rightarrow (p + e) + \bar{\nu} \rightarrow (\nu + \bar{\nu}) + \bar{\nu} = \nu + 2\bar{\nu}$, where $\bar{\nu}$ is an antineutrino. (The first reaction is allowed by low energy weak interaction physics, since it conserves B and L separately. Recall that an antineutrino has lepton number -1.) An alternative reaction would be to use the neutrino from the proton-electron annihilation reaction to give $n + \nu \rightarrow \nu + \bar{\nu}$. Either reaction will result in the conversion of all atoms into neutrinos and antineutrinos. In the first reaction, each neutrino and antineutrino would carry away one-third of the neutron's rest mass or ⅓ GeV (actually somewhat less when we take into account the mass defect of the atom in which the neutron is bound). Since the second reaction is really $(p + e) + n \rightarrow (\nu + \bar{\nu}) + n = (n + \nu) + \bar{\nu} \rightarrow (\nu + \bar{\nu}) + \bar{\nu} = \nu + 2\bar{\nu}$, where the intermediate particles are virtual, each particle in the final state would also have slightly less than ⅓ GeV of energy. Notice that in the first reaction, namely $n \rightarrow \nu + 2\bar{\nu}$, B − L is still conserved. Initially, we have $B - L = +1 - 0 = +1$. In the final state of one neutrino and two antineutrinos, we have $B - L = 0 - ((+1) + 2\,(-1)) = +1$. The net effect of annihilating either a proton-electron pair or a neutron is to reduce the baryon number of the universe by 1 and the lepton number of the universe by 1. As we shall see, this reduction is important to understanding why the Son came to Earth in the first place.

This new law of electroweak baryogenesis has never been seen experimentally in the laboratory, because the energy required for the process to occur at an observable rate is beyond the reach of our particle accelerators. The energy required to overcome the potential barrier between states with different baryon numbers is approximately 10 TeV, or 10,000 GeV. This number is 10 trillion electron volts of energy. The energy available in the Tevatron at Fermilab is about 2 TeV, and it will be about 5 TeV in the Large Hadron Collider now under construction at the European Organization for Nuclear Research (known by its French acronym, CERN). In both of these machines, the energy is divided among the many quarks, antiquarks, and gluons that make up the protons and antiprotons being collided. Only in a very rare event would all the 2 or 5 trillion electron volts be in single particles making up the protons and antiprotons that collide. So we cannot expect to see the electroweak baryogenesis process operating in the immediate future. If this process cannot be seen in a single proton-antiproton collision in our most powerful accelerators—machines that are more than a mile in diameter—how can I expect to see the process act to annihilate the 5×10^{28} atoms that make up Jesus' body?

Electroweak baryogenesis can occur by quantum tunneling through the 10 TeV energy barrier that separates the states of different baryon numbers. Quantum tunneling is observed in many quantum systems, and it works by a cooperative effort across the worlds of the multiverse. Suppose a particle has an energy of only 1 eV and we want it to surmount an energy barrier of 10 TeV. This is impossible if only one universe exists, because that particle has only 1 eV of energy. But the particle is actually spread out over many universes, in each of which it has 1 eV of energy. If each version of this particle in 20 trillion universes provides half of its energy to a single one of the versions, then this single version now has 10 TeV, just enough energy to surmount the barrier. The reason that barrier penetration by this mechanism is never seen in daily life is that it requires cooperation between the worlds of the multiverse. Further, the amount of cooperation required is proportional to the height of the energy barrier, relative to the average energy of each version of the particle. If the barrier is 10 TeV high and each particle has only a single electron volt of energy, then at least 10 trillion particles have to transfer their energy across the worlds to a single particle. The probability of this occurring is exceedingly low under normal circumstances. We see the phenomenon of quantum tunneling—in transistors, for example—only when the height of the barrier is small relative to the energy of each version and only when it is easy to maintain quantum coherence (the possibility of cooperation between the worlds).

I am proposing that the Son and the Father Singularities guided the worlds of the multiverse to concentrate the energy of the particles constituting Jesus in our universe into the Jesus of our universe. In effect, Jesus' dead body, lying in the tomb, would have been enveloped in a sphaleron field. This field would have dematerialized Jesus' body into neutrinos and antineutrinos in a fraction of a second, after which the energy transferred to this world would have been transferred back to the other worlds from whence it came. Reversing this process (by having neutrinos and antineutrinos—almost certainly not the original neutrinos and antineutrinos dematerialized from Jesus' body—materialize into another body) would generate Jesus' Resurrection body.

If a body were to dematerialize via this mechanism inside a linen shroud, it would generate an image just like the image of Jesus seen on the Shroud of Turin. To see this, let us note the key features of the image. First, it is very faint, and in each spot on the Shroud it is limited to the uppermost fibers of a single strand of the linen, as shown in the photograph of a single strand of linen from the Shroud in Figure 8.1. The image is a yellowing of this outermost fiber. Second, the image was formed when the cloth was parallel to the body. That is, the linen was not wrapped around the body when the image was formed. If the cloth had been wrapped around the body when the image was formed, the image would have been distorted. Instead, the image looks like a photograph, which is imprinted on a flat plane. Thus, the image must have been formed when the linen was essentially flat relative to the body. The image-generation mechanism did not affect the linen fibers underneath the blood on the Shroud. Since in addition the blood clots on the Shroud were not deformed, the image-generation mechanism acted without first pulling the linen away from the body in the regions of the

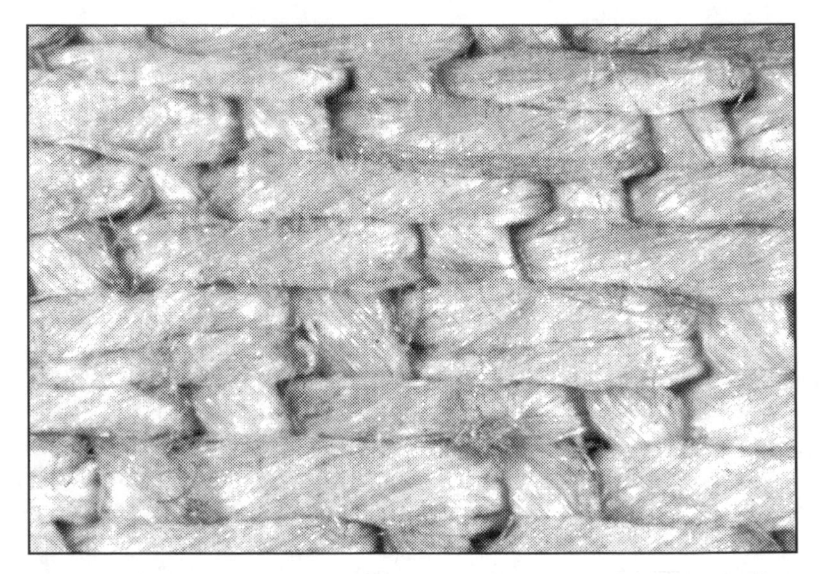

Figure 8.1. Close-up of Turin Shroud body image. The marks are on the very top of the linen fibers.

clotted blood, which would have glued the linen to the body at these places.

But the image is more than a photograph, because it contains three-dimensional information. When a VP-8 analyzer scanned the image, the scan appeared as an undistorted human being in three dimensions. A flat photograph of a human face similarly scanned appears distorted. Alan Adler and John Heller have pointed out that the three-dimensional effect is a consequence of the image contrast on the Shroud being generated by having more fibers yellowed rather than having the same number of fibers become more yellow.[11] The amount of yellowing on each of the image fibers is constant. John Jackson and his coworkers were able to reproduce the features of the Shroud image by putting a bust of a human head that had first been coated with a phosphorescent paint in a tank of water to which ink had been added. When a photograph of this bust was made, it showed the same three-dimensional pattern in the VP-8 analyzer.[12]

The Shroud image could not have been a scorch because it did not fluoresce under ultraviolet, whereas burn marks on the Shroud from a fire in 1532 *did* fluoresce. As a further test, a heated bust of a human head was placed near a sheet of linen. The heat produced an image on the linen, but the nose—which was the closest to the cloth—was burned almost black while other parts of the face did not appear at all.

Chemical analyses of the Shroud image fibers conducted by John Heller and Alan Adler indicate that the image—the yellowing of the fibers—was the result of a conjugated dicarbonyl group formed out of the cellulose of the linen.[13] They were able to undo the yellowing of an image fiber by applying the powerful reductant diimide to the fiber. Upon application of diimide, the image fiber turned white; that is, the yellowing that had formed the image disappeared. Diimide (the compound H-N = N-H) reduces molecules by hydrogenation.[14] That is, diimide adds hydrogen atoms to other molecules. In organic chemistry, *reduction* is defined as a process in which hydrogen atoms are added to a molecule. *Oxidation,* the reverse of reduction, is defined as the removal of hydrogen atoms from a molecule. The image fibers appeared under the microscope to be more degraded (or eaten away) than the nonimage fibers. From these two observations together, Heller and Adler concluded that the image had been formed by oxidation—the removal of hydrogen atoms.

Heller and Adler were also able to yellow modern linen fibers by placing the linen in concentrated sulfuric acid.[15] As is well known, concentrated sulfuric acid is not only a strong acid but also a strong dehydrator. A common experiment in high school is to pour concentrated sulfuric acid into sugar. A black mass rises from the container of the acid and sugar. What has happened is that the sulfuric acid has pulled water molecules (H_20) out of the sugar (glucose $C_6H_{12}O_6$), leaving the carbon. Heller and Adler discovered that dehydration of linen in an alkaline environment failed to produce a yellowing of the linen. Only dehydration in an acid environment produced a yellowing.

I wrote earlier that neutrinos have so little interaction with matter that the mass of Jesus' body could have been converted into neutrinos without the people nearby being affected by the neutrinos. This is true, but the conversion of an 81-kilogram body entirely into neutrinos would have had an effect on the Shroud, which was placed directly on Jesus' body in the tomb. The neutrinos would have had just enough interaction with the atoms of the Shroud to lift the Shroud, exactly the effect required to explain the observation that the Shroud must have been straightened out away from the body just before the image was formed. But the neutrinos would not have exerted sufficient force to pull the

Shroud away from the blood clots, where the linen had become glued to Jesus' body.

The calculation is similar to the calculation which showed that a directed neutrino beam generated by sphaleron action could support Jesus' body. The mass of the Shroud—the object that must be lifted—is obtained as follows. The Shroud measures 437 by 111 centimeters,[16] giving a total area of 48,500 square centimeters (rounding to three-place accuracy). The area density of the Shroud is 22 ± 2 milligrams per square centimeter[17] (the area density was measured to be 21.4 milligrams per square centimeter in the region cut away for the radiocarbon dating).[18] Multiplying the area by the area density gives a total Shroud mass of 1.1 ± 0.1 kilograms. This is the mass that must be lifted. (Or rather half of it, since half of the Shroud would remain under Jesus' body. But I'm just doing an order of magnitude calculation here, because a complete calculation would also have to consider additional forces, such as the force required to pull the Shroud out from under Jesus' body. These forces will be within an order of magnitude of the force required to lift the Shroud.)

A rough order-of-magnitude estimate for the neutrino-antineutrino center-of-mass cross section with neutrons and protons is $\sigma = G_F^2 s$, where G_F is the Fermi coupling constant and s is the square of the system center of mass-energy.[19] This simple relation follows, of course, from dimensional analysis (using particle physics units in which $h/2\pi = c = 1$, so that energy has dimensions of inverse length. The Fermi coupling constant has units of inverse energy squared: $G_F = 1.16637 \times 10^{-5}$ GeV^{-2}). In more conventional units, the cross section is thus $(5.31 \times 10^{-42}\ m^2)(s/\text{GeV}^2)$. This last notation means that we must measure the energy in GeV. From this cross section we can compute the energy and momentum transferred to the Shroud from the neutrinos and antineutrinos as follows.

The body of the man on the Shroud has been estimated by Dr. Robert Bucklin, a forensic pathologist, to have weighed 178 pounds, or 80.8 kilograms,[20] so I shall use this number as our best guess of Jesus' mass. This mass must be converted into neutrinos and antineutrinos. As I noted, a proton-electron pair can be converted into an arbitrary number of neutrino-antineutrino pairs, but let me first do the calculation for the amount of energy absorbed by the Shroud, assuming only one neutrino-antineutrino pair per proton-electron pair. We shall see that this results in too much energy being absorbed by the Shroud, even though neutrinos interact very weakly with matter. There is simply an enormous amount of mass-energy, by nuclear physics standards, in an 80-kilogram man. But I shall show that we can fix the number of neutrino-antineutrino pairs by requiring that no human would incur radiation damage by the act of Jesus dematerializing next to that person.

Recall that the total cross section is defined as the ratio of the total number of interactions to the product of the number of neutrinos-antineutrinos produced via the complete conversion of Jesus' body into neutrinos-antineutrinos and the total number of nucleons (in the Shroud). The number of neutrinos-antineutrinos produced per nucleon is twice what I called N in the reaction $p + e \rightarrow NZ \rightarrow N(\nu + \bar{\nu})$. If we call the total number of interactions N, then the total energy deposited will be EN, where E is the energy per interaction. The number of nucleons in the Shroud equals the total mass of the Shroud in grams (1,100) times the number of nucleons per gram, which is Avogadro's number, 6.022×10^{23}. This gives 6.6×10^{26} nucleons in the Shroud. The number of incident neutrinos equals ($2N$)(mass of a 178-pound man in grams)(6.022×10^{23}) = $N \times 10^{29}$.

For the moment, let me set $N = 1$ to show what will be the result. Then all the energy generated per annihilation of a proton-electron pair will appear in a single neutrino-antineutrino pair, or ½ GeV for each particle. The center of mass-energy squared will be 2 GeV^2 for ½ GeV neutrinos or antineutrinos.

Since s, the center of mass-energy squared, will be seen later to be very important, I will outline the relativity calculation. By definition, $s \equiv (p_n + p_\nu)^2 \equiv (p_n + p_\nu)^\mu (p_n + p_\nu)_\mu = (p_n + p_\nu)_t{}^2 - (p_n + p_\nu)_x{}^2$, where p_n is the 4-momentum of the nucleon and p_ν is the 4-momentum of the neutrino (or antineutrino).[21] In the system of coordinates in which the Shroud is not moving, we have $p_n = (m_n, 0, 0, 0,)$, where m_n is the rest mass of the nucleon (roughly 1 GeV), and $p_\nu = (E_\nu, E_\nu/c, 0, 0)$ is the 4-momentum of the neutrino and antineutrino, where E_ν is the energy of the neutrino or antineutrino. Thus, $s = (E_\nu + m_n)^2 - (E_\nu/c)^2$. Using standard particle physics units in which the speed of light $c = 1$, we see that setting E_ν = (½) GeV, and $M_N = 1$ GeV, gives $s = 2(\text{GeV})^2$. Notice also that if E_ν is very small in comparison with m_n, the value of s essentially becomes the mass of the nucleon squared.

Putting $s = 2(\text{GeV})^2$ gives a cross section of $1.06 \times 10^{-42}\ m^2$, and thus the total number of interactions of the neutrinos or antineutrinos with the Shroud will be 6.6×10^{14}. If the total energy of each neutrino or antineutrino were deposited in the Shroud, as would happen if the collisions between the Shroud nucleons and the neutrinos were perfectly inelastic, the total energy deposited on the Shroud would be (6.6×10^{14} interactions)(½) (GeV per interaction)(10^9 eV/GeV)(1.602×10^{-19} joules per eV) = 53,000 joules. This is an enormous amount of energy. It corresponds to an absorbed radiation dose of (53,000 joules/1.1 kg) × (1 rad/0.01 J/kg) = 5.3 million rads. To see the effect of such a dose of radiation on a human, we have to convert from rads to rems (*rem* stands for "*r*adiation *e*quivalent *m*an"). No human technology has ever generated enough neutrinos to actually endanger a human, so the measurements required to obtain the conversion from rads to rems have never been carried out. However, a reasonable estimate is to assume that the conversion would be midway between gamma rays (for which 1 rad = 1 rem) and neutrons (for which 1 rad = 2 rems). Since I am calculating order of magnitude effects, I shall use 1 rad = 1 rem as the conversion.

Thus, the radiation dosage from the neutrino-antineutrino pairs, assuming $N = 1$, implies a radiation dosage of 5.3 million rems. A lethal dosage of radiation is 600 rems. The recommended maximum dosage for health workers such as X-ray technicians is 5 rems per year. So if Jesus were to dematerialize into ½ GeV neutrinos, anyone standing nearby would receive 10,000 times the lethal dose of radiation. This lethal dose occurs even though a neutrino has the smallest interaction cross section of any known particle. This calculation rules out high-energy ½ GeV neutrinos. In other words, it rules out $N = 1$.

But it does *not* rule out dematerialization into neutrino-antineutrino pairs. I shall now show that dematerialization is possible, with arbitrarily small radiation dosage, if N is sufficiently large. More precisely, I shall show that if $N = 10^6$ or greater, then a person standing next to Jesus when he dematerialized would receive less than 5 rems of radiation. I shall also show that, with $N = 10^6$, there would be just enough momentum transfer from the neutrinos to the Shroud to lift the Shroud.

The key is to realize that with $N = 10^6$, the neutrinos and antineutrinos would have an energy of (½)GeV $\times 10^{-6}$ = 500 eV, or (½) KeV. These are very-low-energy neutrinos, and they will have too low an energy to induce nuclear transitions. They will instead collide elastically with the quarks of the nucleons and the electrons of the atoms. I shall now show that the energy transferred in an elastic collision varies as N^{-2}, while the number of neutrinos increases as N. Thus, the total energy transferred varies as the product of the energy transferred per neutrino times the total number of neutrinos, which is to say, as $1/N$. Thus, we see that by having $N = 10^6$ or higher, the radiation dosage is reduced to 5 rems or lower.[22]

In both elastic and inelastic collisions, the 4-momenta before and after the collision have to be equal. For an elastic collision, the timelike component of the 4-momenta is $E_\nu + m_n = \underline{E}_\nu + \gamma m_n$ and the spacelike component of the 4-momenta is $E_\nu = -\underline{E}_\nu + \gamma m_n v$. Once again, I have set $c = 1$. I have also used the symbol $\underline{E}_\nu$ to denote the energy of the neutrino after the collision with the nucleon, v denotes the velocity of the nucleon after the collision, and $\gamma = (1 - v^2)^{1/2}$ as usual. Adding these two equations and dividing the result by m_n yields $(2E_\nu + m_n)/ m_n = \gamma(1 + v)$. Setting for convenience $A \equiv (2E_\nu + m_n)/ m_n$, we discover (after a bit of algebra) that $\gamma = (1 + A^2)/2A$. What we want to calculate is the kinetic energy KE transferred to each nucleon by the elastic collision with a neutrino. This kinetic energy is $KE = (\gamma - 1)m_n$. But we have $E_\nu = m_n/2N$, so $A = 1 + 1/N$, and hence the $KE = m_n/2(N^2 + N)$. Thus, if $N >> 1$, we will have $KE = m_n/2(N^2)$, which is the formula I wished to obtain. (Strictly speaking, I should use a different coupling constant, call it G_Z, in the cross section in place of the Fermi coupling constant G_F, since the Fermi coupling constant has the W boson in the propagator, and the W boson generates inelastic collisions. It is the Z boson that is responsible for elastic collisions between neutrinos and quarks and electrons. But since $[G_Z/G_F]^2 = [M_W/M_Z]^4 = [80\ \text{GeV}/91\ \text{GeV}]^4 = 0.60$, keeping the Fermi coupling constant is accurate to within an order of magnitude.)[23]

Let me now show that the neutrino flux from dematerialization has just enough momentum to lift the Shroud. Let me first assume that the neutrinos and antineutrinos impart to the Shroud 5 rads of radiation. This amounts to 5.5×10^{-2} joules of energy, so the height this energy can raise the half of the Shroud above the body is $h = KE/mg = 0.51$ cm (half the KE, but half the mass to be raised). This number could be increased inside the tomb, since the intensity of radiation absorbed would drop off as the square of the distance from the body, so if the radiation were to be held at 5 rads (to protect the guards outside the tomb), say 2 meters from the body, the radiation level 1 centimeter from the body could impart $(200\ \text{cm}/1\ \text{cm})^2 \times 5$ rads $= 40{,}000 \times 5$ rads or 200,000 rads (2,000 joules inserted into a 1-kilogram Shroud). John Jackson and

Eric Jumper have estimated that the Shroud was lifted by no more than 4 centimeters by the Resurrection Event.[24] If we set $h = 4$ cm, we obtain 40 rads. But let me keep the 1/2-centimeter lifting ability.

The force exerted on the Shroud depends on the time, Δt, over which the body is dematerialized. The force is given by $F = (2m(KE))^{1/2}/\Delta t = (2(1.1 \text{ kg}/2)(5.5 \times 10^{\ 2} \text{ J}/2))^{1/2}/\Delta t = 0.17$ newton seconds/Δt. If we assume that the dematerialization occurs sufficiently rapidly that it appears instantaneous to a human observer (1/100 second, since this is the response time for the human eye), the force exerted on the Shroud would be 17 newtons. This number should be compared with the 1.1 kg $\times$ 9.8 $m/s^{\ 2}$ = 11 newtons required to cancel the force of gravity on the Shroud. Exerted over the entire Shroud with surface area of 4.85 square meters, 17 newtons would correspond to a pressure of 17 N/4.85 m^2 = 3.5 pascals = 5.1 $\times$ $10^{\ 4}$ pounds per square inch. A small pressure indeed! But the pressure could be raised to 1/10 pound per square inch by the above mentioned possible increase in the energy dosage.

The force exerted on the Shroud would also be increased if the dematerialization time were decreased to the time it takes light to travel across the body. For a body thickness of about 30 centimeters, this would be $\Delta t = 0.30m/3.00 \times 10^8$ m/sec = $(10^{\ 7})(1/100)$ sec. This decrease in the dematerialization time would increase the force and hence the pressure by a factor of 10^7, to 3.5×10^7 pascals, or 5,100 pounds per square inch. In all of these scenarios, I conclude that there will be sufficient energy and force to lift the Shroud, and to raise it sufficiently to make an essentially level surface as observed, as Jackson has shown is required.

But the neutrinos could not generate the image on the Shroud. There might be sufficient energy, but the energy transfer will be uniform throughout the Shroud, because of the low cross section of neutrinos. Nor could neutrinos be responsible for the medieval date obtained by the radiocarbon laboratories. For one thing, the energy of the neutrinos is too low to induce a nuclear transition from carbon 12 to carbon 14. In addition, we would expect that, were the energies sufficiently high to induce such a transition, even more carbon 12 would be converted into carbon 13, and such an increase was not seen.

I propose instead that the sphaleron field itself generated the Shroud image. I suggest that while the Shroud was lifted by the neutrinos, the layers nearest the body were in the sphaleron field, so the atoms of these layers were themselves dematerialized. But we would expect the field to decohere rapidly away from the body, so only the outermost layers would be affected. We would also expect the strength of the field to drop off exponentially away from the body, so the dematerialization effect of the sphaleron field would also drop off exponentially as one goes away from the body. This would explain the three-dimensional pattern in the VP-8 analyzer. The exponential drop-off is exactly the same as the exponential drop-off in light intensity from light passing through an ink medium. (Recall that Jackson showed a light-emitting bust in water to which ink has been added duplicated the three-dimensional pattern in the VP-8 analyzer.) John Heller and Alan Adler observed that the image fibers appeared to be eaten away, and this is exactly the effect of dematerialization.

Dematerializing cellulose means removing (annihilating) hydrogen atoms—by chemical definition, oxidizing the cellulose. It also involves removing (annihilating) water molecules—by chemical definition, dehydrating the cellulose. In both cases, this is exactly what Heller and Adler observed. Finally, these removals of atoms and molecules will occur with the removal (annihilation) of electrons—and by the chemical definition of G. Lewis, an acid is any substance that accepts (removes from its environment) an electron pair.[25] That is, the oxidation and dehydration would happen as if these processes were occurring in an acid environment. In other words, the annihilation would act just like the chemical process of adding concentrated sulfuric acid to the cellulose of the linen.

With one important exception. With an acid to degrade the linen, we would expect heavier elements, such as iron and calcium, that have become implanted in the cellulose of the linen to be left behind by the acid, resulting in an increased concentration of such heavier elements. Heller and Adler found iron and calcium uniformly throughout the Shroud, except for a higher concentration near the edges, where water exposure had increased the heavier elements. A chemical eating away of the image fibers should leave a very slight concentration of iron near the region. This suggests that the image should have been visible in the X-ray images of the Shroud due to this slightly increased iron density. Instead, the image was invisible in the X-ray images. This could be accounted for by insufficient X-ray sensitivity, or it could be caused by annihilation of the iron atoms. This is a possible experiment to test my sphaleron field explanation for the Shroud image.

There is, I admit, one huge gap in my sphaleron-field dematerialization hypothesis. I have not explained exactly how a sphaleron field could be created on a macroscopic scale. I can say only that such a situation can be imagined if—and only if—we can imagine that a coherent quantum state can be generated and maintained on such a scale. In the laboratory, the difficulty with generating and preserving any coherent state is isolating the state from the world around it. It is interesting to note that the Risen Jesus' command to Mary Magdalene in John 20:17, that she was not to touch him, is extraordinarily suggestive of the necessity of isolation for a coherent state. Such a coherent quantum state would be intrinsically nonlocal. This nonlocality would mean that the information from the interior of Jesus' body would be in part also present on the surface of the body. Thus it would account for the observations that the images of the hands on the Shroud look very long and bony, as in an X-ray, and for the appearance of some bony structures in the image of the face.[26] I shall now give a reason why both such situations might indeed be possible and also explain how we might be able to establish that they did indeed occur in the Man of the Shroud.

The Turin Shroud as Holy Grail

The historian Daniel Scavone has brilliantly argued that the Turin Shroud is the source of the Holy Grail legends.[27] The basic idea is that the Grail legends were first written in Western Europe at roughly the same time that the Latin Christians become aware of the Mandylion of Constantinople, an image of Jesus that disappeared during the sack of this city by the Crusaders in 1204. The Mandylion reappears in history as the Shroud of Turin (if indeed the Shroud is genuine).[28] Scavone shows that the description of the Mandylion (or rather its container) matches the description of the Grail in the earliest versions of the Grail legends. Also, the Grail is always associated with Joseph of Arimathea, who provided the tomb for Jesus and presumably the linen burial cloths. (There were two such cloths, according to John 20:7.) After the Resurrection, Joseph of Arimathea would thus have been the owner of the cloths. Also, the Grail is by tradition the receptacle of Jesus' blood. This is exactly what the Turin Shroud, if genuine, actually is. Mark Guscin has made a persuasive case for the Oviedo Cloth to be the other cloth mentioned in John's Gospel.[29] In outline, his argument is that the distributions of bloodstains on both cloths, the Turin Shroud and the Oviedo

Cloth, are similar, and both have type AB blood, a blood type that is rare in the general human population but fairly common among the Jews of Palestine. The Oviedo Cloth is known to have existed since at least A.D. 1000, and a plausible history can be constructed to place it in Palestine in the early first century. So if we define the Holy Grail as the depository of Jesus' blood, then the Oviedo Cloth and the Turin Shroud are together the Holy Grail.

It is interesting to review the key features of the original Grail legends.[30] The earliest story we have about the Grail is an unfinished poem entitled *Perceval: The Story of the Grail,* by Chrétien de Troyes—a name that may be a pseudonym, since it literally means "A Christian from the Town of Troyes." Chrétien himself attributes the story to "a book which the count [Philip of Flanders] gave me."[31] We do not know exactly when Chrétien wrote the story. Since it was unfinished, and tradition tells us that Chrétien died before completing it, it is plausible that he died in the Holy Land, having gone with Count Philip on the First Crusade in 1090. The count died in the Holy Land in 1091, so this date is the best guess for the death of Chrétien de Troyes also. Chrétien de Troyes most likely had begun his poem before then.

Chrétien's story is roughly as follows. A British knight—in some versions of the early Grail writings Sir Gawain and in others Sir Perceval (Parsifal)—visits a castle where the Grail is kept. On the way, he passes through a desolate land. No people are to be seen; the rivers are dry. Near the castle, Sir Gawain meets a fisherman, later revealed as the king of the Grail Castle. The Fisher King is injured in some manner. In the German version of the early legends, the Fisher King has suffered a sword blow in his masculine member.

Upon entering the castle, Sir Gawain sees a sword that has been broken into two pieces and a lance that is perpetually dripping blood. Finally, he sees the Grail. The Fisher King greets Sir Gawain and hands him the sword with the request to mend it. Sir Gawain is unable to do so (not surprising, since he is a knight; repairing a sword is a job for a swordsmith, or at least a blacksmith). The Fisher King is disappointed and tells Sir Gawain that he will fail in his quest. Only one who can rejoin the sword that has been divided in twain can have the Grail. But the king tells Gawain he will answer any question the knight wishes to ask.

Gawain asks after the lance and is told that it is the Lance of Longinus, which pierced the side of Jesus on the Cross. Gawain asks after the sword but, tired from his trip, falls asleep before he hears the answer.

When Gawain awakens, the castle has disappeared. He sees, however, that the rivers are now flowing, and the land is now verdant. He sees people, who say that they both praise and curse him. Praise him, for he has asked of the lance, and this has partially restored the land to health. Curse him because he did not hear the answer to his question about the sword, nor did he ask of the Grail. Had Gawain done so, had he performed the One Deed (rejoin the sword that has been divided in twain) and asked the Three Questions (What is the Lance? What is the Sword? What is the Grail?), the land would have been fully restored.

The key features of this story appear in the most recent retellings of the Grail legend. In Richard Wagner's *Parsifal,* the king of the Grail Castle—located in northern Spain—has been wounded (but by the Lance of Longinus). In the movie *Excalibur,* the sword (Excalibur) is broken in two and mended by supernatural means. The king (Arthur) has been wounded by the sword and healed by drinking from the cup. The Grail knight (Perceval) cannot find the Grail until a question has been answered, but in this version, it is the Grail that asks the question and Perceval who must answer. In the movie *Indiana Jones and the Last Crusade,* the Grail "knight" (Indiana Jones) must answer *three* questions in order to reach the Grail inside the Grail "castle" (inside a mountain, as in *Parsifal*). Jones must also perform a final "deed," guessing which of a collection of cups is the actual Grail. In *all* versions of the Grail legends, the Holy Grail is a talisman of enormous power. The possessor of the Grail is capable of curing illness, granting immortal life, and reversing the desolation of the Earth.

It is interesting that we can take the key features of the Grail legend seriously as applied to the Shroud of Turin and the Oviedo Cloth. The first order of business for Shroud researchers is to rejoin the sword that has been divided in twain. It must be established that the Turin Shroud and the Oviedo Cloth indeed hold the blood of the same man. The Shroud and the Cloth can be rejoined by carrying out a DNA comparison test on the blood on the two linen pieces. If the DNA on the two cloths is from the same individual, then we will have rejoined the two halves last together in Jesus' tomb. By publicizing the similarities between the Turin Shroud and the Oviedo Cloth, Mark Guscin has made the first step in the rejoining.

We next need to conduct the DNA test for a virgin conception of a male, as I described in the previous chapter. It must be established that there are only two distinct X-chromosome gene alleles present in the blood on the Shroud or the Oviedo Cloth, and that there are Y genes from only one individual. The Y genes of course establish maleness (recall that a sword is also a phallic symbol). The two tests together will establish the Virgin Conception. (What is the Sword?)

Longinus of the Lance was a Roman soldier. The stab by this representative of the Empire of Violence into the side of the Prince of Peace is a symbol of evil. According to Christian dogma, Jesus and His mother were without Original Sin. Testing for a modification of the gene on the X chromosome that apparently codes for violent behavior and may also code for bone growth (which I described in Chapter 7) would go far to establish that Jesus and His mother were not inclined to use force the way every other human is. What is the Lance?

The most difficult question is: What is the Grail? That is, we want to learn from the Shroud exactly how a coherent sphaleron field, a field capable of converting matter into energy (neutrinos or photons), was created on the macroscopic scale of a human body. If a study of the Shroud image at the microscopic level could show how this was done, we would thereby learn three enormously important technologies. First, learning how to maintain a coherent quantum state on the scale of a human body would immediately tell us how to manufacture a quantum computer, which is the ultimate computing machine. Second, learning how to convert matter into photons would provide the ultimate energy source during the expanding phase of universal history (gravitational collapse energy is the ultimate energy source, but this source will not be available until the universe has begun to collapse). Third, learning how to create a directed neutrino beam out of matter annihilation would provide the ultimate rocket. If we could learn these things from the Shroud, the Shroud would be a talisman even more powerful than the Grail of the legends!

There is a good reason for thinking that we can in fact learn these three things from a study of the Shroud. I emphasized in Chapter 3 that the universe is currently accelerating. If this acceleration were to continue forever, the laws of physics would be violated, as I also showed. Therefore, the acceleration must stop. Therefore, there must exist a mechanism to stop the acceleration. If the Standard Model of particle physics is correct, then the acceleration can come from only one source: an imbalance between the electroweak vacuum and the positive cosmological constant. I discussed in Chapter 3 why a positive cosmological constant is required by the Standard Model. But if there is a net number of particles over antiparticles—as all observations indicate—and if the excess of particles was created by electroweak baryogenesis—as must be the case if the Standard Model is correct—then the electroweak vacuum cannot be in its absolute minimum. Thus, the positive cosmological constant is not presently canceled out, and so the universe accelerates. But if the particles were to be annihilated with sufficient rapidity by the inverse of the electroweak process that created them, then the acceleration would stop, and the universe would eventually collapse into the Omega Point, preserving the laws of physics. To put it another way, the laws of physics *require* this to happen.

But as I explained earlier, the particles will not be annihilated with the necessary rapidity by the random use of electroweak baryon annihilation. Only a guided use of this sphaleron process will annihilate matter fast enough. Only if our descendants expand out into the universe and make extensive use of this process will the particles be annihilated fast enough. But if our descendants understand how to use the sphaleron process on a small scale, they will do this automatically. They have to act in this way in order to survive, and they have to know about the process in order for the laws of physics to hold for all time.

It is possible that our descendants will learn how to make practical use of electroweak baryon annihilation through their own efforts. Or it may be that we will need some hints on how to develop this process. And the hints are on the Shroud. But the power that comes with the knowledge of the electroweak baryon annihilation process is gigantic. Remember that the power involves the ability to convert 80.8 kilograms of matter into energy almost instantaneously. If the energy were to appear as photons, this would be equivalent to the explosion of a 1,000-megaton bomb. It would not do for us to have the process before we develop a social system that can handle this power. (A similar argument was used by the U.S. government to justify the invasion of Iraq.) It would also have been dangerous to have a man (or a woman) infected by Original Sin to have this power 2,000 years ago. So we will get the power only when we have learned to use it.

This could be an explanation for the error made in 1988 on the radiocarbon dating of the Shroud. What originally convinced me that the Shroud was a fake was the fact that the date obtained was *precisely* that expected if the Shroud were a medieval forgery. The Shroud first appeared in France in 1355, and the Arizona laboratory obtained a radiocarbon date of 1350. It seems incredible that later contamination came in exactly the right amount to give an exactly incorrect date. Unless the contamination was adjusted (by the Father, Son, and Holy Ghost, acting through the laws of physics) to prevent us from starting extensive research on the Shroud too early and thus obtaining the sphaleron process before we were ready for it. Unless, that is, the contamination were a miracle.

Even many Christians often assume that miracles occurred only in the distant past. But if Christianity is true, then we would expect miracles to happen all the time, even today. On January 6, 1945, a German boy had an experience closely resembling Paul's experience on the road to Damascus.[32] It was the most moving experience of his life and was one reason why he became a Christian theologian. Further, he became one of the very few modern theologians to emphasize that Christian belief must be completely rational. Miracles for this theologian must be completely consistent with the laws of physics. Was the religious experience of this sixteen-year-old boy merely a temporary, random glitch in his brain, or could it have been another miracle? This particular German boy in his later capacity as a theologian was largely responsible for reintroducing rationality into Christian theology. This German theologian, Wolfhart Pannenberg, certainly spent fifteen years in a finally successful attempt to persuade an American physicist (me) that Christianity, undiluted Chalcedonian Christianity, might in fact be true and might even be proven to be true by science.

Needless to say, the medieval Holy Grail storytellers never in their wildest dreams imagined that their stories might embody a true prophecy, especially a true prophecy involving an action that could not be carried out without using twenty-first-century physics. So the reader should feel free to regard my discussion of the Turin Shroud as fulfilling a Holy Grail prophecy as merely a fanciful story. Or the reader could recall Dorothy L. Sayers's profound remark on true prophecy:

> Both in real life and in fiction, it is . . . the mark of convincing prophecy to be fulfilled "all wrong"—that is, along lines of potentiality that neither the prophet nor his contemporaries ever foresaw or guessed at. Thus, Virgil's fourth eclogue is a convincing prophecy of Christ just because, not in spite of, the fact that he supposed himself to be writing about somebody quite different.[33]

Testing the Rocks Tests for the Resurrection and the Assumption

I showed in Chapter 7 that, since Jesus and Mary were genetically much closer than is the case in sexual reproduction, if Original Sin were inherited genetically, as St. Augustine claimed, then the absence of "evil" genes in Jesus would likely be accompanied by their absence in Mary. I showed what "evil genes" could mean empirically, and how the Fall could have been a real event in history. If Jesus were capable of dematerialization and miracles because of His genetic structure, then since Mary had the same genes, she also would be capable of dematerialization. If so, we would have a mechanism for the Assumption. Though it is beyond present technology, it is possible in principle to test the Assumption hypothesis, provided the description of Mary's Assumption by St. John of Damascus is accurate. At the Council of Chalcedon in A.D. 451, St. John quoted St. Juvenal, bishop of Jerusalem, as saying that Mary died in the presence of all the Apostles. But when her tomb was opened shortly after her burial (at the request of St. Thomas), it was found to be empty. If she were assumed via the method of dematerialization into neutrinos I have described, some of the neutrinos would have interacted with the rock surrounding her tomb, and a remnant of this interaction could in principle be detectable. The same would be true of the rock surrounding Jesus' tomb. Tradition tells us where the tombs of Mary and Jesus were, so this effect can be looked for in the future, when our technology improves.

The Roman Catholic dogma of the Assumption was defined in 1950 by Pope Pius XII speaking *ex cathedra,* that is, speaking with infallibility. Protestants deny papal infallibility, and indeed deny the Assumption, on the grounds that there is no mention of Mary's Assumption in the New Testament. Pope Benedict XVI, in his autobiography, recounts an episode that shows the difference between Catholics and Protestants in their view of the Bible and the Tradition that interprets the Bible:

> Before Mary's bodily Assumption into Heaven was defined, all theological faculties in the world were consulted for their opinion. Our teachers' answer was emphatically negative. What here became evident was the one-sidedness, not only of the historical, but also of the historicist method in theology. "Tradition" was identified with what could be proved on the basis of texts. Altaner, the

> patrologist from Würzburg . . . had proven in a scientifically persuasive manner that the doctrine of Mary's Assumption into heaven was unknown before the fifth century; this doctrine, therefore, he argued, could not belong to the "apostolic tradition." And this was his conclusion, which my teachers at Munich shared. The argument is compelling if you understand "tradition" strictly as the handing down of fixed formula and texts. This is the position that our teachers represented [and is the Protestant position]. But if you conceive of "tradition" as the living process whereby the Holy Spirit introduces us to the fullness of truth and teaches us how to understand what previously we could not grasp (cf. John 16:12–13), then subsequent "remembering" (cf. John 16:4, for instance) can come to recognize what it has not caught sight of previously and yet was already handed down in the original Word. But such a perspective was still quite unattainable by German theological thought [in the late 1940s]. In 1949, I think, [the theologian] Gottlieb Söhngen held forth passionately against the possibility of this Marian Dogma. . . . Edmund Schlink, a Lutheran expert in systematic theology from Heidelberg, asked Söhngen point-blank: "But what will you do if the dogma is nevertheless defined? Won't you have to turn your back on the Catholic Church?" After reflecting for a moment, Söhngen answered, "If the dogma comes, then I will remember that the Church is wiser than I and that I must trust her more than my own erudition."[34]

If there is indeed no written record of the Assumption before the fifth century, then we must suppose there was an oral tradition, as described, among the Christians at Jerusalem, a tradition that was, surprisingly, not picked up by writers from outside Jerusalem. We would have to suppose this tradition was purely oral for four centuries, which seems implausible. Or we would have to assume that, as Benedict XVI suggests, the Holy Spirit revealed the truth to Christians in the fifth century. A direct experimental confirmation of the Assumption would therefore test the Catholic claim that the Holy Spirit guides the Church to the realization of important truths, sometimes centuries after the events occurred.

How can we test for the Assumption? If Mary dematerialized as Jesus did, *and the neutrinos all had very low energy,* then there is no test. However, if some—or all—of the neutrinos had high energy, in the MeV range, say, then there is a possible test. Some of the neutrinos, even in

Jesus' case, may have been in this range, for the basic process, the conversion of a proton into a neutrino, would yield a GeV neutrino if the neutrino were single rather than multiple. A neutrino in the MeV energy range, if it were to interact with an atom in rock, would cause the rock atom to recoil, and as this recoiling atom moved through the rock, it would leave a track. This track would remain in the rock surrounding Mary's tomb in Jerusalem, the exterior of which is pictured in Figure 8.2.

Figure 8.2. Photograph of the exterior of the Tomb of Mary, in the Kidron Valley below the Mount of Olives.

We could expect to see the same in the rock surrounding Jesus' tomb, the Holy Sepulcher, the doorway to which is pictured in Figure 8.3.

The techniques for seeing such tracks in rocks were developed in the 1960s, and there is an extensive literature on them.[35] Ancient rocks have radioactive nuclei embedded in them, and over time these decay, producing tracks like those in Figure 8.4.

The key feature to note in figure 8.4 is that the tracks are oriented at random with respect to one another. Nuclear tracks generated by neutrinos produced by Jesus' Resurrection or Mary's Assumption would come from a single direction, that of Jesus' or Mary's body relative to the rocks. This process would generate tracks of a completely different appearance, illustrated in Figure 8.5.

So the experiment to test for Mary's Assumption (and an alternative test of Jesus' Resurrection) is to examine the rocks for nuclear particle

Figure 8.3. Entrance to the Holy Sepulcher, a Jerusalem church built over the site of Jesus' crucifixion and his tomb.

tracks like those of Figure 8.5. Finding such tracks would show that there was an isolated source with MeV energy. If such a signature were found, the next step would be to see if the direction of the isolated source was Mary's tomb or the Holy Sepulcher, respectively.

The stone in the vicinity of Jerusalem contains small particles of ara-

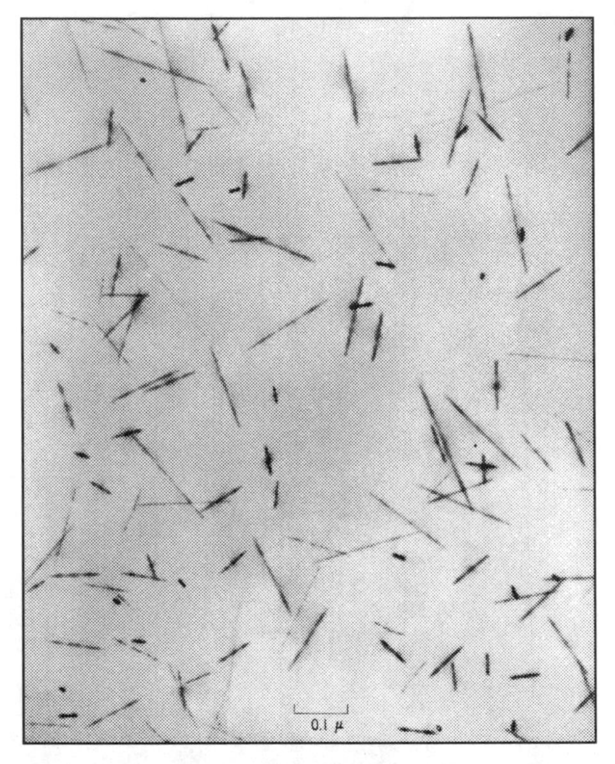

Figure 8.4. Photograph of random nuclear particle tracks.

gonite and calcite.[36] Nuclear particle tracks have been detected on both of these minerals.[37] The test I describe here can be carried out by anyone. There are commercial laboratories that will carry out searches for nuclear particle tracks (these tracks give information about petroleum deposits near the rock). One such firm is Geotrack International in Australia. Robert Fleischer lists three others across the world,[38] but I could not find these other firms on the Internet. They may have gone out of business. There are also university laboratories that specialize in nuclear track studies. A sample of rock could be sent to any of these laboratories, and they could be asked to look for nuclear particle tracks in the aragonite crystals in the rocks. The laboratories should be asked to look for evidence of collimated (parallel, as in Figure 8.5) nuclear tracks rather than random oriented, as in Figure 8.4. If such tracks are found, more extensive investigations should be done.

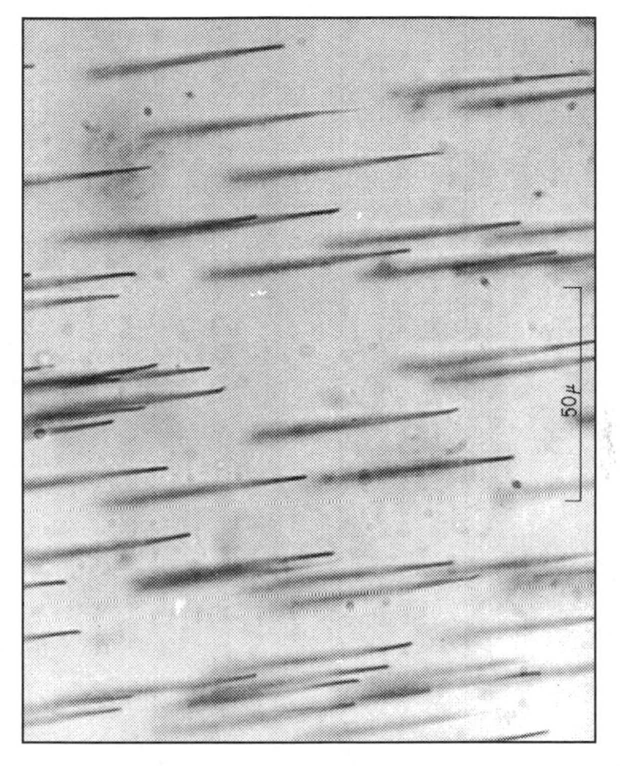

Figure 8.5. Photograph of single-source nuclear particle tracks. If neutrinos of high energy came during Jesus' Resurrection or during Mary's Assumption, we would expect to see tracks like these in the aragonite or calcite crystals in the rocks near Jesus' tomb and around Mary's tomb.

The stone around Jerusalem is a form of limestone called dolomite.[39] In the area of the Holy Sepulcher, it is a soft form of dolomite, called malaki (melekeh), or royal stone, probably the stone used to make Herod's Jerusalem Temple.[40] In the area of Mary's tomb, at the foot of the Kidron Valley (also called the Valley of Jehoshaphat in the Bible), the dolomite is a harder form called mizzi. Both forms of dolomite contain aragonite crystals. Farther up the Mount of Olives, the limestone is a different sort, called cake stone (kakule), containing layers of flint.

The Church of the Holy Sepulcher is a medieval structure built on the site of a church constructed in 326–335 by Helen, the mother of the Roman emperor Constantine, supposedly on the site of both Jesus' cruci-

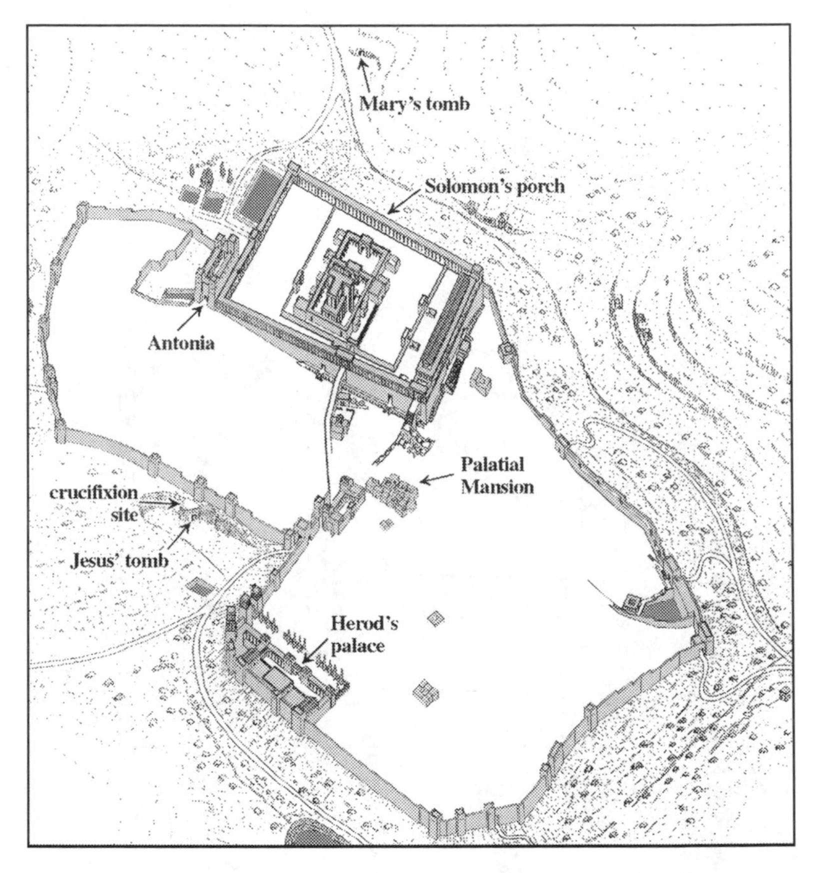

Figure 8.6. Map of Jerusalem in A.D. 30, the year of Jesus' crucifixion and Resurrection. Jesus' tomb is just outside the walls, just under the crucifixion site. Mary's tomb is located on the other side of the city, in the Kidron Valley. Also shown are Solomon's Porch in the Temple complex, where Jesus announced Himself to be God; the Antonia, where Jesus appeared before Pontius Pilate; the Palatial Mansion, where Jesus appeared before the high priest; and Herod's palace.

fixion and his burial.[41] The location of the church inside the walls of Jerusalem will confuse modern readers, because the Gospels make clear the crucifixion and burial were outside the walls of the city. But there is no inconsistency if one keeps in mind the fact that the present-day walls are not the walls of 2,000 years ago. A drawing of the walls in A.D. 30, the year of Jesus' crucifixion (see Chapter 6), is given in Figure 8.6. The crucifixion and burial sites are just above and in the quarry just outside the walls of the ancient city, respectively. This is a quarry of royal stone, located so Herod did not have to transport the stone for his temple very far.[42]

The quarry was brought within the walls in A.D. 43. According to the tradition of the Jerusalem Christian community, celebrations were held on the site until the beginning of the Jewish revolt against Rome in 66. There was another, bloodier, Jewish revolt in the early 130s, and once again the Romans won the war. But this time the Roman emperor, Hadrian, decided to replace Jerusalem with a Greek city. Thus, the Romans covered the quarry in 135 so that a pagan temple of Aphrodite would have a level base.[43] In September 1009, the anti-Christian muslim Caliph al-Hakim ordered the razing of Constantine's Church of the Holy Sepulcher.[44] A Muslim writer, Yahya ibn Sa'id, described the destruction as total:

> They dismantled the Church of the Resurrection to its very foundations, apart from what could not be destroyed or pulled up, and they also destroyed the Golgotha and the Church of St. Constantine and all that they contained, as well as the sacred grave stones. They even tried to dig up the graves and wipe out all traces of their existence. Indeed they broke and uprooted most of them. . . . The authorities took all the other property belonging to the Church of the Holy Sepulcher and its pious foundations and all its furnishings and treasures.[45]

So the original tomb is no longer in existence. Fortunately, Caliph al-Hakim was unable to obliterate the bedrock upon which the Church of the Holy Sepulcher rested, so it is still possible to carry out an experiment on this material.

Taking rock samples from the tomb itself has not been possible for a thousand years. However, it must be said that Constantine's engineers did not realize rock samples from near the tomb would prove useful to physicists seventeen centuries later, and they themselves removed a vast amount of rock from around the tomb in their construction of the original basilica. A pity, because if high-energy tracks are indeed found in the rock near the Holy Sepulcher, we would expect the density of these tracks to decrease as the distance squared from the location of the tomb.[46]

IX

The Grand Christian Miracle: The Incarnation

> The high priest asked him . . . Art thou the Christ, the Son of the Blessed? And Jesus said, I am: and ye shall see the Son of Man sitting on the right hand of power, and coming in the clouds of heaven.
>
> MARK 14:61–62 (KJV)

> In the beginning was the Word, and the Word was with God, and the Word was God.
>
> JOHN 1:1 (KJV)

> Jesus said unto them . . . before Abraham was born, I am.
>
> JOHN 8:58 (KJV)

> I and My Father are one.
>
> JOHN 10:30 (KJV)

C. S. LEWIS CALLED THE INCARNATION "THE GRAND MIRACLE." A "miracle" is a "wonder," an event completely contrary to expectation, and the central claim of Christianity, that the man Jesus is also God, the Second Person of the Triune God, is certainly counter-intuitive. The high priest and many other Jews of 2,000 years ago thought Jesus' claim to be God to be blasphemy. When Jesus asserted that he existed before Abraham was born—and worse, claimed for himself God's Name from Exodus 3:14—many of his Jewish listeners wanted to stone him for blasphemy. For those ancient listeners and most of humanity today, it was and is obviously impossible for a man to be God.

It *is* possible for a man to claim to be God. A contemporary of Jesus, the Roman emperor Caligula, claimed to be a god, an equal of the Roman gods Jupiter and Neptune. Everyone from the first century onward has regarded Caligula as a lunatic. In fact, contemporaries of Caligula record many acts of madness on his part; the claim of godhood was just one. But the picture of Jesus we have from the Gospels indicates total sanity—except possibly for his godhood claim.

So, was Jesus sane? Is it possible to be both God and man? In this chapter, I shall show how the known and extensively tested laws of physics could allow a human being actually to be God or, more precisely, one of the three Hypostases of the Cosmological Singularity.

A Scientific Explanation of the Incarnation

A summary of the orthodox view of how Jesus could be both God and man can be found in the Athanasian Creed, which I have included in the appendix. Jesus is declared to have two natures, one human and one divine. He derives his human substance from his mother, Mary, and his divine substance from his Father, God. He has two wills, one human and capable of sin (though Jesus in fact never sinned) and one divine and hence necessarily infallible. The two wills and the two natures are united by "oneness of person." It is the orthodox view of the Incarnation that is allowed by modern physics.

The key idea needed to understand how the Incarnation could work is the multiverse. Recall from Chapter 2 that quantum mechanics demonstrates that reality consists not merely of a single universe but instead of an uncountable number of universes exactly like our own, an uncountable number of universes more or less like our own, and finally, an uncountable number of universes quite unlike our own. This huge collection of universes comprises the multiverse. We normal humans have one crucial property in common across the multiverse: we and our analogues can exist only in the universes that are either exactly like our own or more or less like our own. Nothing like us can exist in any universe that is quite unlike our own. In particular, nothing human can exist in any universe that remains close to the All-Presents Singularity, since such universes are very small throughout their entire histories. If a universe's maximum size is only an inch across, it will be very difficult to fit a human being in that universe!

This means that analogues of us are restricted to a finite region of the multiverse. (The technical word for "finite" in this context is *compact.*) The restriction of us and our analogues to a compact region of the multiverse is one way we can be regarded as creatures: as part of the created order rather than the uncreated reality that is the Cosmological Singularity. Another limitation we have is our unawareness of the various analogues of ourselves in the multiverse. We are aware of only one universe and of only one version of ourselves, namely, the particular unique humans that are ourselves in this particular universe.

But recall that this unawareness of the other universes is not a fundamental feature of physics. It is only a feature of the design of human brains. David Deutsch showed decades ago that a mentality based on a quantum computer *could* be aware of at least some of the analogues of itself across the multiverse.[1] Let us suppose that such a mentality exists, and that its awareness extends across the multiverse all the way into the All-Presents Singularity. In contrast to a human being, a quantum-computer mentality is not restricted in form and may be of arbitrarily small size if suitably constructed. Recall that the laws of physics require the existence of computers of arbitrarily smaller and smaller size to exist and continue to exist as the universe collapses into the Ultimate Future Singularity.

A mentality that (1) has analogues of itself that exist all the way into the All-Presents Singularity and (2) is aware of these analogues would be a completely different sort of entity than we are and even fundamentally different from those computers of arbitrarily small size that will one day exist near the Ultimate Future Singularity. The computers near the Ultimate Future Singularity will not be aware of the alternate versions of themselves, just as we are not aware of the alternate versions of ourselves in the multiverse. In contrast, a mentality with the two properties I just mentioned would be simultaneously aware of its analogues. Furthermore, this awareness would be transferred not only across the universes of the multiverse but also across the multiverse directly from the All-Presents Singularity itself!

Moreover, the set of all analogues of this hypothetical mentality would *be* the All-Presents Singularity. Recall the mathematical construction in Chapter 4 whereby we were able to identify a portion of the Cosmological Singularity with a set of an infinite number of points inside space and time, or inside the multiverse. Any set of points in either spacetime or the multiverse that approached arbitrarily close to the Singularity, the set being regarded as a single entity, would *be* the Singularity.

If Jesus were to have analogues throughout the entire multiverse going arbitrarily close to the All-Presents Singularity, this set of analogues, regarded as a single entity (recall that "being regarded as a single entity" is just what the word *set* means), then this set of Jesus' analogues would, as a matter of mathematical fact, *be* the All-Presents Singularity. The set of analogues would actually be God or, more precisely, the Second Hypostasis of the single Cosmological Singularity.

Since it is essential for this identification that the analogues go all the way into the Second Hypostasis Singularity, no other human being could be so identified. We and our analogues cover only a finite region of the multiverse, and hence our analogues do not approach arbitrarily close to any singularity. But there is a deeper reason why only Jesus would be God. We are not aware of the other analogues of ourselves in the other universes of the multiverse. This lack of awareness separates us from our analogues. But by hypothesis, Jesus was aware of His analogues, and this awareness crossed the multiverse all the way into the Second Hypostasis of the Cosmological Singularity. This awareness in effect would extend from the Second Hypostasis to the Jesus of our universe. This awareness would convey instructions (the technical mathematical term is *boundary conditions)* from the Second Hypostasis to the man Jesus in our universe of the multiverse. These instructions would be the Will of the Second Hypostasis, which would be distinct from the will of the man Jesus in our universe. Thus, there would be two wills associated with Jesus, the ordinary human will anchored to our universe and the Will expressed across the multiverse, the direct Will of the Second Hypostasis. These two wills together are both expressed in the Man Jesus, and the unity of the two would be the Second Person.

Modern psychology has demonstrated that a similar integration of distinct "wills" into a single person is what happens inside a single human brain in our universe. We may think we have a single will, but in reality, there are different wills in our heads that are molded together to yield a single consciousness, a single person. When the brain's unification mechanism fails, multiple personalities appear; to an outside observer, it is as if there are several people inhabiting the same human body. The human personality is made up of many programs being run on this wet computer called the brain, and there is an integration program that welds them into a single awareness. If this integration program is defective, there is no integration, and multiple personalities develop. The awareness of all of Jesus' analogues across the multiverse, this awareness integrated by the guidance (consistency boundary conditions) of the Second Hypostasis, generates a single Second Person.

Recall from Chapter 4 how distinct hypostases of the Cosmological Singularity were defined: a set of points in a single universe without limit in the future was identified with the Final, or Ultimate Future, Singularity, and another set of points in the same universe without limit in the future was identified with the Initial, or Ultimate Past, Singularity. A third set, comprising a sequence of points in an array of universes, was identified with the All-Presents Singularity. There was an ambiguity in the construction in Chapter 4, because I did not completely spell out the set of points used in the approach to the All-Presents Singularity, and I did not show that the sequence of points in each universe of the multiverse actually defined the same singularity.

These problems can be solved now that we have another entity in the multiverse to define a sequence of points, namely a Person. For the All-Presents Singularity, consider a sequence made up not of spacetime points but of analogues of Jesus in the various universes of the multiverse. If these analogues go all the way into the All-Presents Singularity—and by hypothesis, they do—then the set of all such analogues would *be* the All-Presents Singularity. Recall from Chapter 4 that a set of anything is distinct from the elements defining the set; the set is that collection of elements considered as a unity. Since by hypothesis the analogues of Jesus can act as one, this unity can be called a Person, because this is exactly what it is. Thus, using this sequence of analogues of Jesus to define the All-Presents Singularity, we see that the All-Presents Singularity is a Person, and traditionally it is termed the Second Person. The sequence of the analogues of Jesus going all the way into the All-Presents Singularity is illustrated in Figure 9.1.

A set formed of a sequence of the instantaneous states of an immortal person who goes all the way into the Final Singularity can similarly be used to define the Final Singularity, provided that the knowledge possessed by this person increases to infinity and that, at the Final Singularity, the Person knows everything that can be known. Notice that only at the Final Singularity itself will total knowledge be achieved. At any instant inside spacetime itself, any person will possess only a finite

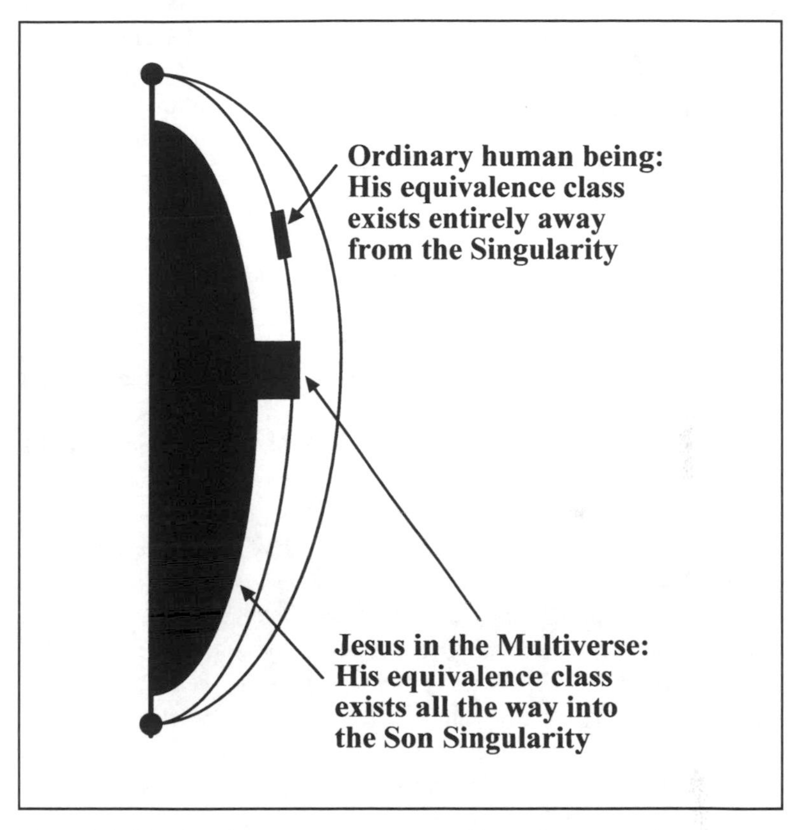

Figure 9.1. The Cauchy sequence, constructed from the person of Jesus across *all* the universes of the multiverse, goes all the way into the Son Hypostasis, and is thus identical to it. Jesus the Man is God by oneness of person, exactly as asserted in the Athanasian Creed.

amount of knowledge. Since the state of total knowledge will necessarily be the same in all universes of the multiverse, the Final Singularity defined by such a personal identification will be identical, so the Final Singularity will be one hypostasis and not many. This is the same unity we obtained in Chapter 4 using only the spacetime points in the superspace of the multiverse, only now we are able to see that the Final Singularity is a Person, distinct from the Second Person. Traditionally, the Final Singularity is termed the First Person.

That is, the Final Singularity—the Ultimate Future Singularity—is identified with God the Father. Recall once again the answer that God the Father gave Moses when Moses asked the former for His name at the Burning Bush: "I SHALL BE WHAT I SHALL BE." God the Father asserted Himself to be future tense. Since God is also ultimate, He must therefore be the Ultimate Future tense, which physics tells us is the Final Singularity. Thinking of causation as acting backward in time, thinking of God as the Final Cause, we see (refer to Figure 4.3, showing the three Hypostases of the Cosmological Singularity) that the Second Person Singularity is united to, is of the same (singularity) substance as, the First Person Singularity. The backward causation suggests the word *begotten* for the relationship between the two hypostases.

The Third Hypostasis is the Initial, or Ultimate Past Singularity. Once again, thinking of backward causation inside the Cosmological Singularity itself, we see that the Ultimate Past Singularity *proceeds* from the Father (First Person) and the Son (Second Person). So it is appropriate to term the Ultimate Past Singularity as God the Holy Spirit, the Third Person. Personhood is appropriate for the Third Hypostasis also because, thinking now of the usual past-to-future causation, it is the Third Hypostasis that talks to us and inspires us, as in Acts 2:4: "And they were all filled with the Holy Spirit, and began to speak in tongues." According to Genesis 1:2, the Holy Spirit—the Third Person—was that Hypostasis present in the beginning of time: "And the Spirit of God hovered over Nothingness." Once again thinking in terms of past-to-future causation, it was the Ultimate Past Singularity that caused the Virgin Birth, as described in Chapter 7. Luke 1:34–35 asserts that the Holy Spirit is the cause of the Virgin Birth: "Then said Mary unto the angel, how shall this be, seeing that I know not a man? And the angel answered and said unto her, the Holy Spirit shall come upon thee, and the power of the Highest shall overshadow thee: therefore also that holy thing which shall be born shall be called the Son of God."

But the All-Presents Singularity was also present at the beginning of time, as seen in Figure 9.1. The opening three verses of the Gospel according to John describe this relationship precisely: "In the beginning was the Word, and the Word was with God, and the Word was God. The same was in the beginning with God. All things were made by him, and without him was not anything made that was made." We only have to identify "Word" with the Second Person.

I shall now relabel the three Hypostases of the Singularity. The Ultimate Past Hypostasis of the Cosmological Singularity is more appropriately called the Holy Spirit, or the Third Person of the Trinity. The

All-Presents Hypostasis of the Cosmological Singularity is more appropriately called the Son, or the Second Person of the Trinity. The Ultimate Future Hypostasis of the Cosmological Singularity is more appropriately called the Father, or the First Person of the Trinity. The relabeled Singularity is pictured in Figure 9.2.

There are passages in the Old Testament that sound as if God had a body like that of a human being. For example,

> Genesis 3:8: They [Adam and Eve] heard the voice of Yahweh Elohim walking in the garden [of Eden] in the cool of the day.

> Genesis 32:24, 26–28, 30: Then Jacob was left alone, and a man wrestled with him until daybreak. . . . Then he [the man] said, Let

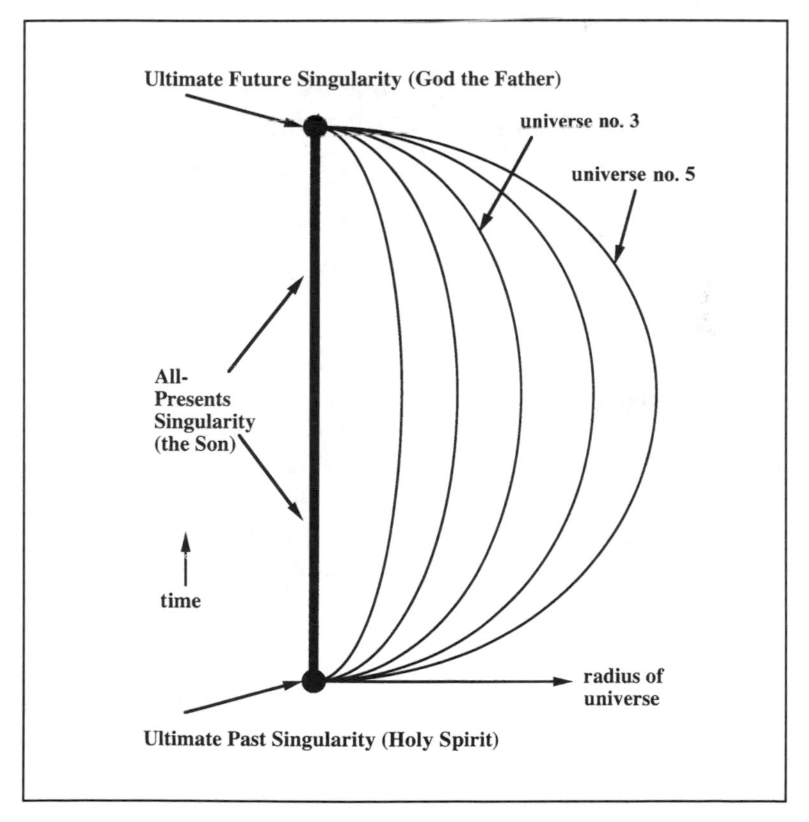

Figure 9.2. The three Persons of the Trinity.

> me go for the dawn is breaking. But he [Jacob] said, I will not let you go unless you bless me. So he [the man] said to him, What is your name? And he said, Jacob. He [the man] said, your name shall no longer be Jacob, but Israel [he who fights with God] for you have striven with God and with men and prevailed. . . . So Jacob named the place Peniel, for he said, I have seen God face to face, yet my life has been preserved.

> Exodus 24:9–10: Then Moses went up with Aaron, Nadab and Abihu, and seventy of the elders of Israel, and they saw the God of Israel; and under His feet there appeared to be a pavement of sapphire, as clear as the sky itself.

> Exodus 33:23: Then I will take My hand away, and you [Moses] shall see My back, but My face shall not be seen.

These passages have generally been given figurative meanings in both Jewish and Christian theology. But if Jesus is God incarnate, then we can imagine that these passages refer to Jesus the man, not to God the Father. Jesus the man was born centuries after the history recounted in these passages occurred, but the incarnation mechanism I have described would allow the man incarnated to appear before he was born. This makes more sense to us than it would have to the ancients, since we have considered constructing a time machine. With such a machine, a man could travel back in time and have a conversation with his great-great-grandfather as a young man.

With the quantum physics of identity, no time machine is necessary. A person identical down to the quantum state would be the person whenever he or she appeared. It is this identity theory that is used in the resurrection theory described in Chapter 3 that allows all of us to be re-created again in the far future. These re-created selves will be us.

There are passages in John that suggest Jesus was the God referred to in the Old Testament passages.

> John 6:46: Not that anyone has seen the Father, except the One who is from God; He has seen the Father.

> John 8:55–58: Yet ye have not known him [Abraham], but I know him, and if I should say, I know him not, I shall be a liar like unto

> you; but I know him, and keep his saying. Your father Abraham rejoiced to see my day, and he saw it and was glad. Then said the Jews unto him, you are not yet fifty years old, and you have seen Abraham? Jesus said unto them, verily, verily I say unto you, before Abraham was, I am.

Jesus claimed to have known Abraham. Justin Martyr, in a book written about A.D. 130 entitled *Dialogue with Trypho, a Jew,* argued that God in these passages was not God the Father but God the Son, Jesus.

The Communion of Saints

The Apostles' Creed asserts a belief in "the communion of saints." This means that the saints, who have died, can nevertheless communicate with us here and now. In the Fatima Apparition, which I described in Chapter 5, three children in 1917 claimed to have spoken with the Virgin Mary, who had a message for the world. In the Roman Catholic Hail Mary prayer, intercession is requested of the Virgin Mary. Such communication between long-dead saints and people living today is possible if the Incarnation theory described in this chapter and the Universal Resurrection theory described in Chapter 3 are both true.

According to the Universal Resurrection theory, everyone, in particular the long-dead saints, will be brought back into existence as computer emulations in the far future, near the Final Singularity, also called God the Father. Communication that is entirely limited to within spacetime is restricted to contemporaries or to those who exist in the future of the senders of the message. This is the communication with which we are familiar. But the Three Hypostases of the Singularity are not restricted to causation acting only from past to future. Future-to-past causation is usual with the Cosmological Singularity. A prayer made today can be transferred by the Singularity to a resurrected saint—the Virgin Mary, say—after the Universal Resurrection. The saint can then reflect on the prayer and, by means of the Son Singularity acting through the multiverse, reply. The reply, via future-to-past causation, is heard before it is made. It is heard billions of years before it is made.

Real Presence in the Eucharist

In Roman Catholic and Greek Orthodox theology, Jesus is really present in the bread and wine after the priest performs the ceremony of the Mass. In this section, I shall show that this notion of Real Presence makes sense as physics. Real Presence is due to the same mechanism that allows a man also to be God. I shall outline an experiment to show that indeed the bread and wine are different from ordinary bread and wine. They have, in a sense, been incarnated also. The Catholic Church has a ceremony called Eucharistic Adoration, in which a person worships a consecrated piece of bread, the host. I shall describe how this makes sense. Several denominations of Christianity deny Real Presence, regarding the Last Supper ceremony as purely symbolic. For example, the Anglican Church—the official Church of England—defines its basic beliefs in the Thirty-nine Articles. Before the end of the nineteenth century, an Anglican clergyman could not be appointed to a living unless he subscribed, in writing, to the Thirty-nine Articles. Article 31 describes the miracle of the Mass—the transubstantiation of the bread and wine into the substance of Jesus—as a "blasphemous fable" and a "dangerous deceit." Testing for Real Presence would be an experimental way of deciding which denomination of Christianity is closer to truth.

First, let us recall the biblical basis of Real Presence.

> "I am the living bread that came down out of heaven; if anyone eats of this bread, he will live forever; and the bread also which I will give for the life of the world is My flesh." Then the Jews began to argue with one another, saying, "How can this man give us His flesh to eat?" So Jesus said to them, "Truly, truly, I say to you, unless you eat the flesh of the Son of Man and drink His blood, you have no life in yourselves. He who eats My flesh and drinks My blood has eternal life, and I will raise him up on the last day. For My flesh is true food, and My blood is true drink. He who eats My flesh and drinks My blood abides in Me, and I in him. As the living Father sent Me, and I live because of the Father, so he who eats Me, he also will live because of Me. This is the bread which came down out of heaven; not as the fathers ate and died; he who eats this bread will live forever." These things He said in the synagogue as He taught in Capernaum. Therefore many of His disciples,

when they heard this, said, "This is a difficult statement; who can listen to it?" But Jesus, conscious that His disciples grumbled at this, said to them, "Does this cause you to stumble? What then if you see the Son of Man ascending to where He was before? It is the Spirit who gives life; the flesh profits nothing; the words that I have spoken to you are spirit and are life."

John 6:51–63 (New American Standard Bible)

Since there is one bread, we who are many are one body; for we all partake of the one bread.

1 Corinthians 10:17 (New American Standard Bible)

Pope John Paul II described the Eucharist as follows:

For over a half century, every day, beginning on 2 November 1946, when I celebrated my first Mass in the Crypt of Saint Leonard in Wawel Cathedral in Krakow, my eyes have gazed in recollection upon the host and the chalice, *where time and space in some way "merge"* [my emphasis] and the drama of Golgotha is re-presented in a living way, thus revealing its mysterious "contemporaneity."[2]

The Eucharist, while commemorating the passion and resurrection, is also in continuity with the incarnation. At the Annunciation Mary conceived the Son of God in the physical reality of his body and blood, thus anticipating within herself what to some degree happens sacramentally in every believer who receives, under the signs of bread and wine, the Lord's body and blood. . . . When, at the Visitation, she bore in her womb the Word made flesh, she became in some way a "tabernacle"—the first "tabernacle" in history—in which the Son of God, still invisible to our human gaze, allowed himself to be adored by Elizabeth, radiating his light as it were through the eyes and the voice of Mary.[3]

The sacramental re-presentation of Christ's sacrifice, crowned by the resurrection, in the Mass involves a most special presence which—in the words of Paul VI—"is called 'real' not as a way of excluding all other types of presence as if they were 'not real,' but because it is a presence in the fullest sense: a substantial presence whereby Christ, the God-Man, is wholly and entirely present."

> This sets forth once more the perennially valid teaching of the Council of Trent: "the consecration of the bread and wine effects the change of the whole substance of the bread into the substance of the body of Christ our Lord, and of the whole substance of the wine into the substance of his blood. And the holy Catholic Church has fittingly and properly called this change transubstantiation." . . . "Do not see—Saint Cyril of Jerusalem exhorts—in the bread and wine merely natural elements, because the Lord has expressly said that they are his body and his blood: faith assures you of this, though your senses suggest otherwise."[4]

And Pope Benedict XVI says,

> In the Eucharist, Christ is really present among us. His presence is not static. It is a dynamic presence, which makes us his, he assimilates us to himself. Augustine understood this very well. Coming from a Platonic formation, it was difficult for him to accept the "incarnate" dimension of Christianity. In particular, he reacted before the prospect of the "Eucharistic meal," which seemed to him unworthy of God. In ordinary meals man becomes stronger, as it is he who assimilates the food, making it an element of his own corporal reality. Only later did Augustine understand that in the Eucharist the exact opposite occurs: the center is Christ who attracts us to himself; he makes us come out of ourselves to make us one with him (cf. *Confessions,* VII, 10, 16). In this way, he introduces us into the community of brothers.[5]

What does it mean to say Jesus is actually present in the bread after the priest performs a certain ceremony? With the Incarnation theory I have described, we can say that, just as there was a coherent connection between the man Jesus and the Second Hypostasis of the Singularity, so there is a coherent connection between the bread after the ceremony and the Second Hypostasis of the Singularity. The priest does nothing. The Second Hypostasis establishes the coherence. Jesus the man established the ceremony 2,000 years ago, and in His divine nature—the Second Hypostasis—He establishes coherence between Himself and the bread today. If such coherence were established, the "substance" of the bread and wine would be changed in exactly the same way that the substance of the man Jesus was transformed by coherence across the multiverse into the Second Hypostasis of the Singularity. The atoms of the bread and wine are unchanged as atoms, just the way the substance of the man Jesus was from his mother, the same as ordinary humans. *Transubstantiation* in modern physics terminology is *quantum coherence.*

Seen under a microscope, the consecrated bread looks like bread. Standard chemical tests would show that the consecrated host is indistinguishable from bread. Consecrated bread, under the theory of Real Presence developed here, differs from bread only by being in a coherent quantum state with the Second Hypostasis of the Singularity. As discussed in Chapter 2, an electron may be in a coherent state with another electron, or it may not. Coherence in spacetime is determined by the history of the electron.

If an experiment is performed to see whether an electron is an electron, ignoring the question of coherence, in both cases—an electron in a coherent state, or not—the electron is confirmed to be an electron. Coherence is more subtle, a relationship with other objects rather than something intrinsic to an object. A determination of quantum coherence requires a test on the two electrons that are suspected to be in quantum coherence. Similarly, a test of quantum coherence in a consecrated host would be far more subtle than just tests for intrinsic properties of the bread.

Maintaining quantum coherence between two electrons, or any large system, is exceedingly difficult, as discussed in Chapter 8. Quantum coherence on a large scale is essential in causing matter to be converted into neutrinos. Maintaining quantum coherence on a large scale is the chief barrier to creating a quantum computer. If one electron of a pair of electrons in a coherent state interacts with the exterior environment, coherence is lost. What is different in Real Presence is the action of the Second Hypostasis of the Singularity. It is this action that can maintain coherence against the host's exterior environment. If two molecules of the host were in a coherent state, information about one would provide information about the other. This would not be the case if the two molecules were not in a coherent state. If two electrons were formed in a single coherent state, knowing the spin of one would automatically allow us to know the spin of the other. The Incarnation theory developed here does not specify what coherent state the material of the host would be in. So it is not possible to describe exactly an experiment to establish coherence. But if coherence is suspected, trial and error can establish what form the coherence takes.

Transubstantiation

The Fourth Lateran Council, held in the year A.D. 1215, defined Roman Catholic understanding of the physical nature of Real Presence. Many Protestants do not object to Real Presence as such, but they do object to the explanation of Real Presence as transubstantiation. *Transubstantiation* means that the *substance* of the bread and wine has been transformed into—replaced with—the substance of Jesus. The *accidents*—the appearance of the bread and wine, or more precisely how the bread and wine look to the unaided human senses and even to simple measuring devices—are unchanged.

This doctrine of transubstantiation will be seen to be equivalent to my description of Real Presence as quantum coherence with the Son Singularity, if we first translate the observational physics of the thirteenth century into modern language. First of all, we need to realize what people meant by *substance* at that time. They believed that all objects on the planet Earth were composed of various amounts of four fundamental substances—earth, air, fire, and water. Thus, transubstantiation would occur if any of these fundamental substances were replaced with another, even if the appearance of the material being observed were unchanged.

An example of such transubstantiation—in their eyes, not ours—would occur if the following experiment were carried out. Fill a sealed room with "air," consisting of a mixture (in our view) of hydrogen gas and oxygen gas. If a person were to enter such a room, he or she would find this mixture completely breathable, without any noticeable negative effects, at least if the person did not stay in the room for very long. A mixture of gases (in our view) is "air" by their view, since the air we breathe is actually composed of two main gases, nitrogen and oxygen. "Air" is not fundamental in our view. Now light a match (or make a spark, using flint and steel, since they didn't have matches) in the room. An explosion occurs, and the room is now damp. The "air" in the room has been replaced with water. That is, a tiny amount of one fundamental substance, fire, introduced into another fundamental element, air, has caused the air to be transformed—transubstantiated—into a third fundamental substance, water.

Now let us analyze what has really happened using quantum mechanics. There has been no change in "substance" in our meaning of the word. The number of hydrogen atoms and the number of oxygen atoms is unchanged. There has, however, been a change in the quantum coherence relations between the atoms. Before the water was formed, the hydrogen atoms were bound in pairs to each other to form a hydrogen molecule, and the oxygen atoms were bound in pairs to form an oxygen molecule. These molecules individually were in a coherent quantum state. Further, the atoms were bound in such a molecule precisely to form a coherent quantum state, since being in such a state would minimize energy. But in a mixture of hydrogen and oxygen gas, there is a lower energy coherent state, the state in which the hydrogen and oxygen are bound into water molecules. The mixture would become water eventually even without the spark; the spark merely speeds up the reaction rate.

So what has really happened from the quantum-mechanical point of view is one coherent quantum state has been replaced by another. In other words, what people in the thirteenth century would regard as transubstantiation, we would regard as a change in the quantum-coherence relations between the physical objects.

This is exactly what I am proposing happens when the bread and wine are transubstantiated by the priest. The atoms making up the bread and wine are unchanged. Also, most of the coherence relations between the atoms making up the bread and wine are unchanged, so all simple chemical tests performed on the Host would show nothing but bread and wine. A far more subtle test, capable of showing that the bread and wine have, after transubstantiation, a coherence with the Son Singularity, would demonstrate that Real Presence is a fact of physics.

The Lutheran doctrine on Real Presence is expressed in Article 10 of the *Confessio Augustana,* where it is said that the body and blood of Christ are truly present *(vere adsint)* in the celebration of the meal. The Lutheran leader Philipp Melanchthon (1497–1560) wrote in his book *Apology,* which is also part of the Lutheran confessional writings, that there is no difference concerning this issue between the Lutherans and the Roman Catholic and Greek Orthodox churches. There is an opposition only to the Roman Catholic doctrine on transubstantiation, not concerning the Real Presence. Transubstantiation was considered by Luther and by several other leading reformers as an expression of Real Presence in a specific philosophical language, the language of Aristotelian ontology. There, transubstantiation would be a paradox, since otherwise all change is an accidental property, not in the substance, while accidents remain intact. Thus, many Protestant churches, such as Lutherans, did not reject Real Presence, only transubstantiation as its mechanism.[6]

I hope that by restating the doctrine of transubstantiation from the language of Aristotelian physics into the language of quantum physics, I have made the concept more acceptable to those Protestants who accept Real Presence. This description of Real Presence can also provide a way of establishing by experiment whether Real Presence is an actual phenomenon. It provides a way of testing between those sects of Christianity that believe in Real Presence and those that do not.

But I have done more in this chapter. I have shown that Real Presence is the same sort of phenomenon as the Incarnation. Both connect physical entities in this universe with the Second Person Singularity via quantum coherence across the universes of the multiverse. Both Real Presence and the Incarnation thus make perfect sense in modern physics.

The Incarnation, the claim that a man is also God, is the feature of Christianity that traditional Jews, who otherwise have much in common with Christians, find most objectionable. We will now turn to the relationship between Christianity and Judaism.

X

Anti-Semitism Is Anti-Christian

Now the LORD said to Abram, . . . "And I will bless those who bless you, and the one who curses you I will curse. And in you all the families of the earth will be blessed."

GENESIS 12:1, 3 (NEW AMERICAN STANDARD BIBLE)

And all the people said, "His blood shall be on us and on our children!"

MATTHEW 27:25 (NEW AMERICAN STANDARD BIBLE)

So I say, Has God put his people on one side? Let there be no such thought. For I am of Israel, of the seed of Abraham, of the tribe of Benjamin. God has not put away the people of his selection. . . . For it is my desire, brothers, that this secret may be clear to you, so that you may not have pride in your knowledge, that Israel has been made hard in part, till all the Gentiles have come in; And so all Israel will get salvation. . . . As far as the good news is in question, they are cut off from God on account of you, but as far as the selection is in question, they are loved on account of the fathers. Because God's selection and his mercies may not be changed.

ROMANS 11:1–2, 25–26, 28–29 (BASIC ENGLISH BIBLE)

> I [Paul] am a Jew of Tarsus in Cilicia by birth, but I had my education in this town at the feet of Gamaliel, being trained in the keeping of every detail of the law of our fathers; given up to the cause of God with all my heart, as you are today.
>
> ACTS 22:3 (BASIC ENGLISH BIBLE)

> Let there be no thought that I have come to put an end to the Law or the Prophets. I have not come for destruction, but to make complete. Truly I say to you, Till heaven and earth come to an end, not the smallest letter or part of a letter will in any way be taken from the Law, till all things are done.
>
> MATTHEW 5:17–18 (BASIC ENGLISH BIBLE)

> But heaven and earth will come to an end before the smallest letter of the law may be dropped out.
>
> LUKE 16:17 (BASIC ENGLISH BIBLE)

AS THE PASSAGES OF THE NEW TESTAMENT JUST QUOTED make clear, anti-Semitism was not present in the early Christian Church. Jesus explicitly denied any wish to set aside the Covenant that God made with Abraham, the ultimate ancestor of all Jews.

But regarding salvation, there are passages in the New Testament such as these:

> I am the way, the truth, and the life: no man comes to the Father, but by me.
>
> John 14:6 (Jesus is the speaker)

> There is no salvation in anyone else, for there is no other name in the whole world given to men by which we are to be saved.
>
> Acts 4:12 (Peter is the speaker)

> The gospel was preached also to them that are dead, that they might be judged according to men in the flesh, but live according to God in the spirit.
>
> 1 Peter 4:6

Christians and Jews

Christianity and Judaism are two brothers with the same father, the religion of the God Yahweh of Abraham, of Jacob, and of Moses. They agree on the fundamental Bible—Christians call this basic Bible the Old Testament, while (in English) Jews call it the Hebrew Bible. Both brothers were born in the first century of the common era, what in this book I have termed, in the Christian way, A.D.—*anno Domini,* "in the year of the Lord." In the century of their birth, the two brothers began to disagree on the interpretation of the "basic Bible," and both eventually created written records of the proper interpretation, in their opinions. The Christian book of interpretation is called the New Testament, and the Jewish book of interpretation is called the Talmud.

Christianity and Judaism are two brothers in the same sense that the people of the United States and England are brothers. Both Americans and the English speak the same language, and their culture, both political and intellectual, is descended from the culture of England in the seventeenth and eighteenth centuries. Had both nations not been literate, their languages would have become incomprehensible to each other by now, 200 years after their separation. A thousand years ago, the language of England, Anglo-Saxon, and the language of present-day Germany were mutually comprehensible, but not today: the Anglo-Saxons and their cousins in Saxony were largely illiterate, so the languages diverged. In illiterate cultures, languages change rapidly. After two centuries, six to ten generations, the language of a tribe will have so changed that the forebears would not understand their great-great-great-great-great-great-grandchildren, even though each generation learned their language literally at their parents' feet.

So oral traditions can change over time if the written word does not nail them down. The non-Christian Jews of the first century were the heirs of a rich oral tradition (later to be largely codified in the Talmud). The Christians of the time accused the "Jews" of deviating from the true line of interpretation, and this is a possibility, if an oral tradition, like a language, is not written down. Commentators, like the native speakers of a language, modify the tradition by the act of commenting on it.

The word *Jews* is in quotation marks in the previous paragraph because only a sect of the Jews in the first century, the Pharisees, adhered to the Oral Tradition. The Sadducees, the Jewish priests who oversaw the sacrifices in the Temple in Jerusalem, denied the Oral Tradition and accepted only the authority of the Pentateuch, otherwise known as the Torah or the five books of Moses (the first five books of the Old Testament).[1] The "Jews" of modern times are the religious descendants of only the Pharisees. The Sadducees were wiped out by the Romans in A.D. 70, and the Jewish Christians ceased to think of themselves as Jews by the end of the second century.

It is true that many major Christian leaders have made very definite anti-Semitic remarks over the past two millennia. For example, in 1543, the same year in which Copernicus published his book *On the Revolutions,* wherein he argued that the Earth is not the center of the universe, Martin Luther published a book entitled *On the Jews and Their Lies,* wherein he gave the "sincere advice" to the German princes to burn down the synagogues, confiscate the Jewish prayer books, outlaw religious teaching by rabbis, raze Jewish houses to the ground, and force the Jews to do nothing but manual labor.[2]

But the same Martin Luther had published a book twenty years earlier, in 1523, entitled *That Jesus Christ Was Born a Jew,* wherein he wrote: "If the apostles, who were also Jews, had dealt with us Gentiles as we Gentiles deal with the Jews, there would never have been a Christian among the Gentiles. Since they dealt with us Gentiles in such brotherly fashion, we in turn ought to treat the Jews in a brotherly manner in order that we might convert some of them. . . . We should remember that we are but Gentiles, while the Jews are in the lineage of Christ."[3] The reason for Luther's change of heart concerning the Jews is that they declined to be converted to Christianity. They listened politely to his arguments, to his quotations from the Hebrew Bible, gave counterarguments to his theses, and presented alternative—and to their minds, more persuasive—interpretations of the biblical passages. They all had heard Luther's arguments before. The Christian-Jewish debate had been going on for centuries, and no new argument had appeared in that time.

There has been an astonishing rebirth of anti-Semitism in the first few years of the twenty-first century. I shall discuss only one example of recent anti-Semitism, an appeal delivered in January 2005 to the prosecutor general of the Russian Federation. This example is important for two reasons: first, it was promulgated by educated men, including one famous mathematician who should have known better; and second, it contains some of the older theological objections by Christians against

Judaism. Here are a few excerpts from this appeal, entitled "Jewish Happiness, Russian Tears":

> One of the thirteen main principles of Judaism is waiting for the Jewish ruler of the world [the Messiah], who will install Jews to rule over all other peoples of the world. . . . The Orthodox Church teaches us that this ruler is identified as the anti-Christ, about which Jesus Christ (John 5:43), Paul, and the holy fathers of the Church warned. This is an important part of the Orthodox teachings of the Apocalypse.

The emergence of an evil ruler, who will attempt to conquer the world before the Second Coming of Jesus, is standard doctrine, but needless to say, the identification of this evil ruler, the anti-Christ, with the Messiah is *not* Orthodox doctrine.

> Because the Jewish religion is anti-Christian and misanthropic, some of its more zealous followers practice ritual killings. Many cases of this ritual extremism were also proved in court, [specifically] a court case from 1884, describing the killing of Christian babies by Jews.
>
> The New Testament explains this misanthropy with words of Christ when he talks about Jewish spiritual leaders denying God the Father and crucifying the Son of God: "your father is the devil and you want to carry out his desires; he was a killer of man from the beginning" (John 8:19, 44). This is a widely accepted Orthodox explanation of Jewish aggression as a form of Satanism.[4]

There is no evidence whatsoever that Jews have ever engaged in ritual murder of anyone, much less Christian babies. Citing a czarist Russian court case to the contrary is equivalent to citing the conviction of the Salem witches as evidence of the existence of witchcraft. Jesus did indeed accuse "the Jews" of having the Devil for their father. But this accusation must be put in the context of the time when He was speaking. Jesus was Himself a Jew, as were all of his disciples. He obviously did not regard Himself as a son of the Devil. Jews of the time routinely denounced their opponents as sons of the Devil. In fact, in the same chapter of John, "the Jews" accuse Jesus of being possessed by a devil (John 8:48, 52). The expression "the Jews" appears five times in Matthew, six times in Mark, five times in Luke, and seventy-one times in John.[5] But similar language can be found in the Dead Sea Scrolls;[6] those Jews who defend the majority opinion of the Jewish community are called "the Jews" by other Jews who hold a minority opinion. The Gospels of Matthew and John are traditionally held to have been written by the Apostles Matthew and John, both Jews who were Christians, and hence in the minority, in the first century, when these Gospels were written.

The Jewishness of Matthew must be kept in mind when we interpret the "blood libel" passage of Matthew 27:25: "His blood shall be on us and on our children!" Matthew obviously did not intend it to apply to all Jews, since he himself was a Jew. The particular sect of Jews who made the remark were the Sadducees, and they and their families were all killed by the Roman army in A.D. 70. This blood debt was paid 2,000 years ago.

The Christian Jews may have formed a majority of the Jews by the end of the second century. At the start of the first century, there were an estimated 10 million Jews in the Roman Empire, but by the end of the third century, this number had fallen to only 1 million. What happened to most of the Jews? The Jewish-Roman wars of A.D. 70 and 130 would account for the disappearance of only a couple of million. Excavations have shown that most of the Christian churches in the first and second centuries were located in the Jewish areas of the major cities of the Roman Empire, quite close to synagogues. This means that, in the second century, most Christians may have been former Jews.[7] More precisely, by the end of the second century, the Jewish people had split into two groups, the majority group calling themselves "Christians" and not "Jews." The fact that Christianity was dominated by Jews at the time the Gnostic heresy appeared in the second century may be a reason why Christianity remained true to its Jewish roots.

Conversely, the Gnostic heresy, which I discussed in Chapter 5, may be a major source of anti-Semitism. Recall that in the Gnostic heresy there are two gods, of roughly equal power, one the good god of the spiritual realm, and the other the evil god who created the material universe. The Jews claim their God created the material world, no question about that. The Gnostic heretics are therefore logically forced to regard the Jews as the servants of the evil god. So we would expect to see Gnostic heretics advocating the persecution of Jews. Thus, I would argue that an anti-Semite is under vehement suspicion of heresy. It is interesting that Igor Shafarevich, the Russian mathematician whose work against the Gnostic heresy I cited in Chapter 5, signed the anti-Semitic petition "Jewish Happiness, Russian Tears," from which I just quoted. As was the case with the Dominicans of the Inquisition, often a scholar is captured by the very heresy he warns against.

According to Orthodox Judaism, Jews must observe 613 commandments (the Mitzvot) in order to be worthy of eternal life in heaven. Non-Jews, however, need observe only seven commandments, the Noahide Laws, to earn a place in the afterlife. The first three Laws (no sexual immorality, no murder, and no stealing) and the last two (no eating meat from a live animal and the requirement to establish courts to enforce the Law) are unproblematic from the Christian point of view. The fourth Noahide Law (the prohibition of idolatry) and the fifth (condemning blasphemy) may be fundamentally inconsistent with the Incarnation. I say "may be," because, as I discussed in Chapter 9, there are passages in the Torah which, interpreted literally, say that God has anthropomorphic attributes, and there is a Christian tradition going back to the second century which says that, in these verses, "God" means the man Jesus. However, standard Orthodox Judaism has interpreted these passages figuratively, and has insisted on an absolute distinction between God and man, any man. When Jesus explicitly asserted His divinity, and indeed claimed to have known Abraham personally, as recorded in John 8:55–58, the Jews prepared to stone him. According to Mark 14:61–64, the high priest said Jesus' claim to be God, made before the assembled Sanhedrin court, was "blasphemy." If the high priest spoke for all subsequent Judaism, then no Christian can follow all the Noahide Laws.

Conversely, many Christians have interpreted Jesus' words that no one can inherit eternal life except through Him as meaning that all Jews necessarily go to hell. This in fact has been the dominant Christian view since the second century, and this interpretation has been cited as a justification for anti-Semitism. However, in the latter part of the twentieth century, the Roman Catholic Church began to take a more subtle view of these passages. Remembering that all Jews who followed the Law were promised eternal life in the later books of the Old Testament and remembering Jesus' words that the Law will never be suspended, the Church decided that indeed Jews as Jews merit eternal life. Therefore, the Church has discontinued missionary efforts to convert the Jews. Jesus' requirement that one can come to the Father only through Him is fulfilled if Jesus appears to an Orthodox Jew as God in the afterlife, with the result that this Orthodox Jew, even if he was vehemently anti-Christian in this life, comes to believe in Jesus after death. Some of the Christian creeds—for example, the Apostles' Creed reproduced in the Appendix—prefigure this solution when they assert that Jesus "descended into Hell" in the three days between His crucifixion and His Resurrection, in order to save those who were in hell because they did not believe in Him when they died. How could they believe in Him if they died before He was born? Or if they were born in a distant land and had never heard the Gospel preached?

I think the current Catholic interpretation is correct, because I am convinced that God never, ever, changes His word. The Jews are promised heaven if they obey the Law, and heaven they will get. This entire book, based on God's Word, expressed in physical law, is never, ever, set aside. Therefore, anti-Semitism is anti-Christian, and antiscience.

The Second Coming

Christians and Jews are agreed on the event that will settle the issue of which religion, Christianity or Judaism, is the true religion: the Coming of the Messiah. According to Christians, the *future* coming of the Messiah will be the Second Coming of Jesus, at which time He will appear explicitly as the Second Person of the Trinity. Jews, by contrast, believe that the Messiah will be a military leader who will rule over Israel and will be a mere man, someone other than Jesus.

According to the Talmud, specifically tractate Avodah Zarah, the universe will exist for only 6,000 years: the first 2,000 years mankind will not have the Torah, the next 2,000 years will be the period of the Torah (this period was considered to be the time from Abraham to the completion of the Mishna, the first part of the Talmud), and the final 2,000 years will be the period of the Messiah, in other words, the Messiah could come at any time during this period. To put it another way, the Talmud predicts that the Messiah must come by the year 6000 in the Jewish calendar, or the year A.D. 2240 in the standard calendar.

According to the computer technologist Ray Kurtzweil, development of computers is proceeding so fast that we should see the arrival of machines that are our intellectual superiors, and the arrival of the technology that will allow the creation of human downloads, by the year 2042.[8]

According to me, in my 1994 book, *The Physics of Immortality,* we should achieve human-level artificial intelligence by the year 2030 if, as I believed then and still believe now, 10^{15} bits of memory and 10 teraflops of computer speed in the typical laptop computer of the day are required.[9] Kurtzweil is more conservative than I, believing that more memory and speed are required. We are projected to have 10^{17} bits of memory and 100,000 teraflops of speed available for laptops by the year 2037, if Moore's law for the rapidity of development of computer hardware continues to hold. Moore's law says that memory and processing speeds for computers increase by a factor of 100 every seven years, and this rate has been remarkably stable over the past few decades.

In my book with John D. Barrow, *The Anthropic Cosmological Principle,* we recorded that the fastest supercomputer then available, the Cray-2, had a speed of 1 gigaflop (that's 1 billion flops).[10] In *The Physics of Immortality,* I mentioned that the fastest supercomputer in 1992 was a 100-gigaflop machine, and I projected that, by the year 2002, a supercomputer with a speed of 1,000 teraflops should be available.[11] I overestimated: the fastest Japanese (NEC) computer in 2002 had a speed of only 35.8 teraflops,[12] and the Cray X1 had a speed of only 51 teraflops.[13] So supercomputer speeds increased by a mere factor of 500 in ten years, whereas I projected a factor of 10,000. I should have believed Moore's law, which predicted a factor increase of roughly 500 in ten years. The IBM Blue Gene/L, with a speed of 300 teraflops at the end of 2005, comes very close to my 1992 prediction of 1,000 teraflops by 2002.[14]

Achieving 1,000 teraflops in 2006 rather than 2002 means that it took fourteen years rather than ten, or about 40 percent longer than I expected in 1992. But the upshot is, supercomputers now exceed my 1992 estimate for the equivalent computing power of the human brain. *We don't lack the hardware for human-level artificial intelligence; we lack the software.* We still don't have a clue how the human-soul program works. Developing the artificial-intelligence software will require a vast number of independent researchers, and this in turn means that the 10-teraflop speed has to make it to the laptop level. Historically, laptop speed has lagged supercomputer speed by twenty years. So human brain–level laptops should be on everyone's desk by 2025. Human-level artificial intelligence will take some time to develop, and I'm still sticking to my 2030 prediction.

According to the Christian Evangelical James Rutz, conversions to Christianity are increasing at an exponential rate, so fast in fact that, at the present rate of increase, every man and woman on Earth will be Christian by the year 2032.[15] More scholarly analyses of the rapid rate of Christian conversions, notably by the sociologist of religion Rodney Stark and the professor of religion Philip Jenkins, yield slower projections, but they still believe that by the end of the twenty-first century, a substantial majority of the human race will be Christian if present trends continue.[16]

One of the reasons this rapid rate of growth in the number of Christians is hard to accept is the fact that most people have an incorrect idea of how Christianity has developed in the United States. Most think that the number of Christians was large in the early years of the American republic and has been steadily decreasing ever since. The opposite has actually occurred. Rodney Stark has carefully studied the U.S. census data, which until recently recorded which church each American belonged to.[17] In 1776 only 17 percent of the American people belonged to any church. By 1850 this had increased to 34 percent. Not until 1906 were a majority of Americans churched. The number of Christians in the United States, Stark discovered, has steadily increased until by 1980, 62 percent of Americans were churched, and the vast majority of Americans attending church attended a Christian denomination. So America started basically as a nonreligious country and has slowly, over its 200-year history, been taken over by Christianity. It is no accident that, in 1980, all three major candidates for the presidency of the United States—John Anderson, Jimmy Carter, and Ronald Reagan—described themselves as Evangelical Christians. I have no doubt that all three men were sincere in their self-descriptions. All three men were typical Americans in their religious beliefs. Past presidents have been much more skeptical. John Adams and Thomas Jefferson were Unitarians, privately scoffing at the idea that Jesus was God, and Abraham Lincoln refused to answer questions about his religious beliefs.

The Christian takeover of America has been slow by historical standards. Stark estimates that the rate of increase in the number of Christians in the Roman Empire was roughly 40 percent per decade.[18] Assuming 1,000 Christians in A.D. 40 (ten years after Jesus' crucifixion), there would have been 6.3 million Christians, or about 10 percent of the population, by the year 300, and 33.9 million, or 56 percent of the population, by the year 350. When the emperor Constantine announced himself to be a Christian in 313, he was embracing what was soon to be the majority faith of the empire.

The twentieth-century rise of Christianity in China has been even more phenomenal. When the Communists took over China in 1949, there were an estimated 700,000 Protestants and 3 to 4 million Roman Catholics.[19] Fifty years later there were between 80 and 100 million Protestants and about 12 million Catholics.[20] Thus, Christians constituted about 7 percent of China's population of 1.26 billion in the year 2000.[21] The number of Protestants in China increased by a factor of 100 in fifty years. If the numbers continue to increase at this rate, a majority of Chinese will be Christian within twenty to thirty years, and the Chinese Christian leaders expect to achieve majority status in China within this time frame.[22] Chinese Protestants have begun what they call the "Back to Jerusalem Project," which aims to convert all people living between China and Jerusalem to Christianity, and they intend to send out at least 100,000 missionaries from China in order to achieve this goal by the time China herself is Christianized.[23] Given their previous success in China, the Chinese Christians may achieve their goal, in which case Rutz's estimate of 2032 as the year the entire world becomes Christian will be accurate.

If Jesus indeed rose from the dead using the mechanism described in Chapter 8, namely electroweak tunneling to convert matter into energy, and if indeed this was done with the intention of showing us how to use the same process, then we ourselves should be able to learn how to turn matter into either electromagnetic energy or neutrinos within a few decades. Converting matter into electromagnetic energy would provide an effectively unlimited energy source, making all present sources of energy—such as coal, oil, and natural gas—obsolete. Converting matter into beamed neutrinos (as walking on water requires) would provide the ideal propulsion system, making all current transportation vehicles obsolete. But remember that if indeed Jesus rose by converting matter into energy, it would also be possible to release the energy all at once. Recall I calculated that a human body's worth of mass converted into energy all at once would constitute the equivalent of a 1,000-megaton bomb. In other words, humans would have the ability to make 1,000-megaton bombs in a home workshop.

In summary, by the year 2050 at the latest, we will see:

1. Intelligent machines more intelligent than humans
2. Human downloads, effectively invulnerable and far more capable than normal humans

3. Most of humanity Christian
4. Effectively unlimited energy
5. A rocket capable of interstellar travel
6. Bombs that are to atomic bombs as atomic bombs are to spitballs, and these weapons will be possessed by practically anybody who wants one.

Furthermore, all of these events and capabilities will occur at very close to the same time. Not only will we and our nonhuman descendants possess a practically unlimited capability for violence, but we will also simultaneously possess the capability to carry this enormous potential for evil into interstellar space. Since only a small minority of the human race needs to be evil for the violence of the weapons I describe to be unleashed, we seem to be destined to see within fifty years the End of Days as described by Jesus:

> For then shall be the Great Tribulation, such as was not since the beginning of the world, no, nor ever shall be. And except those days be shortened, there should be no flesh saved; but for the elect's sake those days shall be shortened. . . . Immediately after the Tribulation of those days shall the Sun be darkened, and the Moon shall not give her light, and the stars shall fall from heaven, and the powers of the heavens shall be shaken. And then shall appear the sign of the Son of Man in heaven: and then shall all the tribes of the Earth mourn, and they shall see the Son of Man coming in the clouds of heaven with power and great glory.
>
> (Matthew 24:21–22, 29–30)

Indeed, the simultaneous appearance of nonhuman intelligence (AIs and human downloads), explosives gigantically more powerful than nuclear, and interstellar propulsion systems will definitely bring on a Great Tribulation for humankind. Furthermore, if this process is not interrupted, it is highly probable that no flesh (human, animal, or plant) should be saved. If the evil in the Earth's biosphere is not to be carried outward to the stars, then a direct intervention of God (the Cosmological Singularity), acting, of course, only through the laws of physics, will be required. I have given reasons in *The Physics of Immortality* why it is unlikely human beings will ever traverse interstellar space, which is the natural environment of AIs and human downloads. We should not be surprised at this; we were informed in the Tower of Babel story (Genesis 11:1–9) that God will not allow humans to inhabit the heavens (except as downloads, which are physically equivalent to our resurrected selves, as discussed in Chapter 3).

So the experiment that will definitely decide between Christianity and Judaism, the identity of the future Messiah, will be carried out within fifty years. We cannot be more precise on the date because the preceding estimates of the rate of technological growth are necessarily imprecise. We will not know the date of the End of Days until they are almost on us. Once again, Jesus' words tell us this: "But of that day and hour knoweth no man, no, not the angels of heaven, but my Father only. . . . Watch therefore: for ye know not what hour your Lord doth come. . . . Therefore be ye also ready: for in such an hour as ye think not the Son of Man cometh" (Matthew 24:36, 42, 44).

Christians (and skeptics) are justifiably suspicious of predictions of the Second Coming, because such predictions have been made in the past, and they have always been proven wrong. In the United States, the most famous prediction of the Second Coming was made by William Miller (1782–1849), who concluded that Jesus would return on October 22, 1844. Miller obtained his date from the book of Ezra (7:12–26), where King Artaxerxes of Persia commanded Ezra to rebuild the temple. Miller inferred that the Second Coming would occur 2,300 years after this command, and he used Bishop James Ussher's date of 457 B.C. for the date the Persian king gave the command. Miller did not realize at first that the standard B.C./A.D. system contains no year 0, so he originally predicted a date of 1843. He himself was unsure about the actual day—he of course knew Jesus' statement that no one knows the actual day. Miller's followers were responsible for October 22.

Notice that Miller's predicted date, as are all other predicted dates before and after, is based on an interpretation of a biblical passage. My prediction is based on physics and the rate of technological change. From physics an event is predicted in the mid-twenty-first century that bears an uncanny resemblance to Jesus' description of His Second Coming. My prediction of the day of the Second Coming is thus unique among such predictions because it comes from well-known physical law and not from my personal interpretation of any biblical passage. This prediction is no different from a prediction, using standard Newtonian mechanics, of the date an asteroid spotted coming toward Earth would collide with our planet.

The Jewish and Christian views of the Coming of the Messiah have several important things in common. One of these is the importance of Jews when the Messiah comes. According to Revelation 7:4–8, at the time of the Second Coming, many Jews will convert to Christianity and play an important role in converting many unbelievers. This is now happening, as the existence of an organization called Jews for Jesus attests. The recently retired Cardinal Archbishop of Paris was a Jew. The influence of contemporary Jews on this book should be obvious, and I have thus dedicated this book to those Jews who are advancing the Christian cause. To hate the Jews is to hate both Christianity and science.

XI

The Problem of Evil and Free Will

But Peter made answer and said to him, though all may be turned away from you, I will never be turned away. Jesus said to him, Truly I say to you that this night, before the hour of the cock's cry, you will say three times that you have no knowledge of me.

MATTHEW 26:33–34 (BASIC ENGLISH BIBLE)

Now Peter was seated in the open square outside the house: and a servant-girl came to him, saying, you were with Jesus the Galilaean. But he said before them all that it was false, saying, I have no knowledge of what you say. And when he had gone out into the doorway, another saw him and said to those who were there, this man was with Jesus the Nazarene. And again he said with an oath, I have no knowledge of the man. And after a little time those who were near came and said to Peter, Truly you are one of them; because your talk is witness against you. Then with curses and oaths he said, I have no knowledge of the man. And straight away there came the cry of a cock. And the word of Jesus came back to Peter, when he said, Before the hour of the cock's cry, you will say three times that you have no knowledge of me. And he went out, weeping bitterly.

MATTHEW 26:69–75 (BASIC ENGLISH BIBLE)

The Lord giveth, and the Lord taketh away. Blessed be the name of the Lord.

JOB 1:21 (TRADITIONAL TRANSLATION)

The Problem of Evil

On December 26, 2004, an earthquake measuring 9.15 on the Richter scale (35 gigatons of seismic energy released) caused a tsunami in the Far East that killed more than 300,000 people.[1] Why did God permit this tragic loss of life to occur? The Christian Church calls this question the "mystery of evil" because it seems that a God Who loves us as a Father and Who is also all-powerful and all-knowing would not allow this horror. Ordinary human fathers in many cases had to watch their children die in the disaster because they were powerless to save them once the tsunami hit. And of course they did not know there were going to be an earthquake and tsunami that day; had they known, they would have kept their children (and themselves) far, far away from the shore.

On August 29, 2005, Hurricane Katrina hit New Orleans. I am a professor at Tulane University in New Orleans, so this was for me a more personal disaster than the Asian tsunami. My family and I evacuated, and by God's grace we suffered little damage to our house, though I had to finish this book as an evacuee under somewhat difficult conditions.[2] But many people did not expect the city to flood, stayed in the city, and died. Once again, why did God permit this tragic loss of life, to say nothing of the loss of property?

God is not powerless, and He does not lack foreknowledge. It seems He could have stopped the tsunami had He wished, or at least warned us had He wished. The question is then, Why did He not wish, if He indeed loves us as a Father? Many people resolve this conundrum by concluding that God does not exist. Indeed, this Problem of Evil is the main cause of atheism not only today but throughout history. If God does exist, there can be only one reason why He allows evil to exist: it is logically impossible for Him to remove evil from reality. Even an omnipotent God cannot do what is logically impossible. Explaining why removing evil from the cosmos is logically impossible is the subject of a discipline called *theodicy*.

All solutions to the Problem of Evil must necessarily be based on some theory of what the physical universe is actually like. All solutions to the Problem of Evil have been based on a false theory of the universe: they have assumed that only a single universe, rather than a multiverse of universes, exists. Let us consider the effect of the existence of the multiverse on the Problem of Evil.

In the multiverse, all universes consistent with physical law actually exist. All human-caused evils could have been avoided, because human decisions to act one way or another are made only after the constraints of physical law are taken into account. There was no physical law forcing Adolf Hitler to murder 6 million Jews. Hitler could have chosen not to kill the Jews, and he could have chosen not to establish the Nazi dictatorship over Germany. Therefore, there is a universe in which Hitler never rose to power and in which the Holocaust never occurred. There is a universe in which none of the evil empires that have murdered people in our universe ever arose. There is no evidence that any human-caused evil is required by physical law. Therefore, there is at least one universe in which no human evil ever occurred. I argued in Chapter 7 that evil arose on Earth much earlier than the human species, that it was present by the time of the Cambrian Explosion. But we can apply the same argument of choice to nonhuman beings and conclude that there must be a universe in which no evil caused by living beings ever arose.

The tsunami of 2004, however, was not caused by human or other living beings. It is an example of a *natural evil*, caused by the automatic operation of physical law. However, in this particular case, we now know that the technology existed to warn most of those who would die from it that the tsunami was coming. Due to human error in this universe, the warning system was not in place. But there must be a universe in the multiverse in which this warning system was in place, with the consequence that very few people died. I argued in Chapter 7 that there is no physical law which prevents backup copies of metazoans from existing. If backup copies of metazoans exist, there need be no evil at all. Certainly death would not exist in a universe in which backup copies of all metazoans existed. Since physical law allows such a universe to exist, it must exist somewhere in the multiverse. Somewhere in the multiverse, there must actually exist the best of all possible worlds. But we obviously don't live in such a universe. The question is then, Why didn't God limit His creation to this best of all possible universes, which does in fact exist?

The answer to this question is connected to the answer to an even more fundamental question: Why did God—the Cosmological Singularity—create even one universe, much less the multiverse? The traditional answer of theologians has been: God's love. We are told in 1 John 4:8 and 1 John 4:16 that "God is Love." The Greek word translated "love" in these passages is *agape,* which means "selfless love." But selfless love by definition requires an object outside the lover; God must by His very nature have an object for His love. Thus, He created the universe to have an object for His love. The historian of ideas Arthur O. Lovejoy has pointed out that theologians of centuries past drew another conclusion about the creative implications of God's love: not only did He create *some* creatures but He created all possible creatures that could fit in the universe.[3] To have done anything less would have meant that His love would lack a possible recipient, and God would not rest until all possible recipients of His love had been created. The philosophers and theologians of centuries past called the list of all creatures that could exist the *Great Chain of Being*. There could be, they argued, no gaps in the chain. All possible ecological niches (to use modern terminology) were filled. All possible plants, from the simplest to the most complex, existed. All possible animals, from single-celled animals, through insects, reptiles, mammals, and primates to human beings existed. And the Great Chain of Being did not stop with humans. There were angels above us in the chain and ranks of angels all the way up to God.

The Great Chain of Being fell out of favor with philosophers in the eighteenth century, once it was realized that there were gaps in the chain in the present epoch and that there were possible animals and plants which had never existed and will never exist in our finite universe.

But in a multiverse of universes there is no reason why all possible animals and plants do not exist in some universe at some time. The existence of the multiverse brings back, indeed requires, a Great Chain of Being that is completely filled within the entire multiverse. A given time in a single particular universe will have gaps. However, in the entire multiverse, each possible species and, further, each possible individual will have his or her chance at existence. The love of God requires nothing less. Theologians have been unable to solve the Problem of Evil because they have not followed the logic of God's love to its ultimate conclusion: God will create until all possible creatures actually exist. The unlimited love of God requires nothing less. They have underestimated what God could do because they have underestimated what He has done. He has created a multiverse, not a universe. We physicists have discovered the multiverse by experiment, but its existence follows necessarily from God's existence and the fact that He is indeed all-good. His all-goodness follows from His love.

Let me put God's limitless love in human terms. We can all observe parents who have children with disabilities. Most of these parents love their disabled children with a love that is equal to the love they bear their other children. Similarly, God loves the disabled universes—ours, for example—just as He loves the universes in which evil never appeared. And His love is manifested in the fact that He creates not only the perfect universes but also the imperfect universes. Similarly, in our imperfect universe, He creates not only human beings with no physical defects but also those with physical defects. He commands us to love both equally, just as He does.

Christianity has always taught that the physically defective are to be loved as much as the physically perfect. This is why Christianity has always been opposed to abortion and infanticide, and opposed the death penalty except in cases where it can be shown that the only way to prevent the later murder of an innocent human being is the execution of a guilty person. In the past few decades, many have rejected the Christian teaching against abortion on the grounds that a fetus is not a human being. This position also rejects the Christian claim that human life begins at conception. The Christian view has always been that the human soul is infused into the fetus at conception and has denied that a fetus is a mere mass of human flesh—a person's tonsils or appendix, for example—whose killing would not be murder. Everyone is agreed that a doctor's removal of a person's appendix or her tonsils would not be murder, and correctly so. The question is whether a fetus is like an adult human or like an appendix. The Christian claim that the fetus is completely human from the instant of conception has been denounced as "unscientific," as an expression of faith—that is, opinion backed up by no facts—rather than reason. But I have argued throughout this book that the key claims of Christianity are science, not faith. The statement that a fetus is a person from the instant of conception is also a statement of science.

Throughout this book I have used the Turing-test definition of *person.* I have used this definition to show that certain very complex computer programs would have to be considered people, even if these computer programs are definitely not of human form. I have also used the Turing-test definition to show that God—the Cosmological Singularity—is a Person or, more precisely, that the three Hypostases of the Cosmological Singularity are Three Persons. As I emphasized in Chapter 3, this definition is based on physics; it has no faith content at all. The Turing test is the standard definition of personhood used by all atheist computer theorists.

The Turing test has a crucial ambiguity, however: namely, it does not specify what sort of conversation the entity being tested for personhood has to have with the normal human being. In my previous applications of the Turing test, I have not specified the necessary conversation in detail; instead, I have allowed *any* conversation the skeptic wishes. This lack of specification means that an artificial-intelligence computer program would have to be much more complex than any human being actually is if the poor program is to pass the Turing test. An AI program would have to be superhuman in order to be counted as human! If I were to require that, in order to count as human, an entity had to carry on a conversation about physics at the level of a full professor of physics, practically no actual human being would pass the Turing test.

David Deutsch, an Oxford University professor of physics, has actually proposed requiring a computer program to have accepted for publication a single-authored paper in *Physical Review Letters* (the leading American physics journal) in order for the program to count as a person. Deutsch and I have both passed the Deutsch test (naturally!), but relatively few actual humans have passed, or ever will pass, the Deutsch test for personhood. However, the reason practically no actual human has passed the Deutsch test is not that the average human cannot even in principle think about physics at a very high level but that the average person has decided not to spend the years of study needed to acquire the skill to think about physics at this level. Most people have taken at most one physics course in either high school or college. They have other interests.

But suppose an evil ruler captured you and gave you a choice: write a paper acceptable for *Physical Review Letters* or be tortured to death. Suppose he was willing to pay to support you for ten years while you studied the necessary math and physics, and also was willing to pay for the best tutors to prepare you for writing the paper. Suppose he also added a carrot to the stick: once the paper is accepted for publication, you will be given a billion dollars and never bothered again. Could you pass the Deutsch test after ten years? I claim that it is likely you could. In spite of what a nonphysicist may think, writing the required paper is not really that difficult. No deep creative thought is necessary; the central ideas in many physics papers, even in *Physical Review Letters,* have been suggested by others, and the authors have merely developed these ideas. You have a physics tutor, remember? Let us suppose the tutor is Albert Einstein. Given the necessary time, resources, and motivation to train, every reader of this book could pass the Deutsch test.

Given less time and less resources than those required to pass the Deutsch test, a fetus, in the process of changing from a fetus to a child, can train itself to have a conversation with an adult human being. Furthermore, the fetus has this power from the moment of conception. We all know this, because we were all once fetuses, and we taught ourselves how to have a normal human conversation after only a decade or two of training, that is, of being normal children. So if we slightly modify the definition of the Turing test to allow an artificial-intelligence program time to reprogram itself if necessary to have the required conversation, a human fetus would be able to pass this modified Turing test also.

The current fierce debate over abortion is more than a debate over whether a fetus is a person. More fundamentally, it is a debate over the solution to the Problem of Evil, the central problem of theology, the central problem of reality, since the solution hinges on realizing that the multiverse exists. And that God really and truly believes all possibilities should be realized. Most abortions currently performed in the United States are actually performed not to eliminate defective children but as a method of birth control—to eliminate inconvenient children.[4] By His actions, God has shown that putting up with inconvenience for the sake of the greater number, be it universes or children, is the morally just action. Allowing evil to come into existence is hugely inconvenient from the divine point of view, because it would necessitate His Incarnation, death, and Resurrection. Dying on the Cross is far more inconvenient than any financial sacrifice required to raise a few extra children. But God the Father knew before time began that the Cross would be necessary in a multiverse of maximum potential, which necessarily means allowing evil to exist.

We have underestimated the extent of God's creation; we did not realize the multiverse existed because we have underestimated His love. The existence of evil is a problem for us only because we have not truly believed that when Jesus said He loves even the most abject sinner, He means it.

God's Foreknowledge and Human Free Will

How can human beings have free will if God foresees everything they will do? Christianity has always asserted that an omniscient God indeed knows what we will do before we do it, yet humans have free will. How is it possible to have both foreknowledge and free will? In physics terms, unitarity is a strong form of determinism. The current state of the multiverse is completely determined not only by its past state but also by its future state. If everything is determined, how is free will possible?

Free will and determinism (God's omniscience) are possible and mutually consistent because reality is a multiverse, not a mere universe, and this multiverse is subject to the quantum theory of identity. God knows whatever it is logically possible to know. He does not know what it is logically impossible to know. Christian theology has always accepted that God cannot do what it is logically impossible to do. God cannot, for example, make a stone so heavy that it would be impossible for Him, an omnipotent being, to lift. Skeptics try to argue that if God cannot make such a stone, He cannot be omnipotent, thus trying to establish a logical contradiction, but there is no contradiction. A stone so heavy that an omnipotent being cannot lift it doesn't mean anything. Such a stone would be like a four-sided triangle. The expression "four-sided triangle" appears to mean something, but it doesn't. A triangle, by definition, has only three sides, and no object can have exactly three sides *and* exactly four sides. God cannot make a four-sided triangle either, but His inability reflects not a limit on God but just a limit on human language. Human language allows us to utter nonsense collections of words such as "four-sided triangle." One of the discoveries of twentieth-century logic is that there can exist unknowable truths. This is one way of expressing Gödel's theorem.

God has constructed reality so that He can both know everything that can be known *and* yield free will for His creatures in this reality. He has created an infinite number of each human being, and He knows what each will do. If a person has a choice between two actions, one good and one evil, God knows that a certain percentage will choose good and a certain percentage will choose evil. He does not know, because it is unknowable, which choice the particular you in this universe of the multiverse will make. The quantum theory of identity makes knowing this particular truth logically impossible. Before the choice is made, there is no difference at the quantum level between the various yous in the universes of the multiverse. It therefore makes no sense to say, before the choice is made, that a particular you will make a particular choice. The only knowable fact allowed by physics is that a certain percentage of you will choose good and a certain percentage bad. The particular you indeed has a free choice. You know that you had a free choice to choose a bad act because a certain percentage of you do in fact choose the bad act. If an event occurs, and was set in the beginning to occur, it is therefore possible to do. Your basic nature will determine what the percentage choosing bad is, and God knows this. And He knew it before the multiverse began.

The Problem of Evil and the problem of free will versus determinism have the same solution: the multiverse, and the quantum mechanics that controls it.

XII

Conclusion

What I say to you in the darkness, speak in the light; what you hear whispered, proclaim on the housetops.

MATTHEW 10:27

Go ye therefore, and teach all nations.

MATTHEW 28:19 (KJV)

And he said unto them, Go ye into all the world, and preach the gospel to every creature.

MARK 16:15 (KJV)

And that repentance and remission of sins should be preached in his name among all nations, beginning at Jerusalem.

LUKE 24:47 (KJV)

But Thomas, one of the twelve, called Didymus, was not with them when Jesus came. The other disciples therefore said unto him, we have seen the Lord. But he said unto them, except I shall see in his hands the print of the nails, and put my finger into the print of the nails, and thrust my hand into his side, I will not believe. And after eight days again his disciples were within, and Thomas with them: then came Jesus, the doors being shut, and stood in the midst, and said, Peace be unto you. Then saith he to Thomas, reach hither thy finger, and behold my hands; and reach hither thy hand, and thrust it into my side: and be not faithless, but believing. And Thomas answered and said unto

> him, My Lord and my God. Jesus saith unto him, Thomas, because thou hast seen me, thou hast believed: blessed are they that have not seen, and yet have believed.
>
> JOHN 20:24–29 (KJV)

IT SHOULD BE OBVIOUS NOW, TO EVEN A CURSORY READER OF this book, that I have been taking Christianity very seriously indeed. I have been treating Christianity not as a "religion"—in the sense this word is often used these days, as a synonym for a collection of myths that no educated person could take seriously—but as a possibly true theory of physical reality. In particular, I have seriously considered the possibility that, as a matter of historical and physical fact, the birth of Jesus was indeed the consequence of a virginal conception and that Jesus may indeed have risen from the dead, just as described in the Gospels. I have shown exactly how both of these miracles could have occurred in a manner that would not violate known physical law.

But taking Christianity seriously as a possibly true scientific theory means that we have to be open to the possibility that it may more closely mirror reality than any other human "religion." A reader uncomfortable with this possibility has been very uncomfortable with this entire book. But I have assumed the truth of Christianity in order to investigate the physical consequences of this assumption. In science, it is impossible to develop the consequences of a theory unless it is tacitly assumed the theory is true. By knowing the consequences of a theory, it becomes possible to propose tests of the theory. Conversely, we have to consider the possibility, in the words of the knights in the *Song of Roland*:

> *Paiens unt tort e Chrestiens unt dreit.*
>
> Pagans are wrong, and Christians are right.[1]

Christianity is right (or may be right) in the sense that any physical theory is right. It may be right in the sense that it is right that the Earth is 4.6 billion years old. It is wrong that the Earth is 6,000 years old, and it is wrong that the Earth is infinitely old. But if Christianity is right in this most fundamental sense of right, then some Christians are wrong in the way they think about God, about the afterlife, about Jesus. Many contemporary Christians have separated the world of faith from the world of science. This divorce allows them to be comfortable with their own particular beliefs and other people's quite different beliefs. But this is a view of Christianity that has been in existence for only the past two centuries. The first Christians had a quite different attitude. The first Christians believed that Jesus really did, as a matter of physical fact, rise from the dead. They also believed that, as matter of physical fact, Jesus had no human father but was instead born in a Virgin Birth. In this book I have shown how the Virgin Birth and the Resurrection could have occurred in ways consistent with known physical law and why these two events may have *had* to occur when they did in history.

The first Christians *really* believed that Jesus was the Son of God and that He rose from the dead. They showed they really believed by being willing to die, if necessary, for their convictions. A number of people who have read an earlier version of this book have asked me if I *really* believe the arguments I am presenting here.

I do. I think of myself as a Physics Fundamentalist, by which I mean that we have to accept as true the consequences of the five fundamental physical laws—quantum mechanics, the Second Law of Thermodynamics, general relativity, quantum cosmology, and the Standard Model—unless and until an experiment shows these laws to have a limited range of applicability. To date all experiments are consistent with these fundamental laws. Therefore, I believe them. Therefore, I believe their consequences, which I have developed in this book. I will continue to believe in the fundamental laws of physics even if doing so results in my professional death as a physicist. It is not acceptable today for a physicist as physicist to believe in God. But I do; I believe in the Cosmological Singularity, which *is* God. I have a salary at Tulane some 40 percent lower than the average for a full professor at Tulane as a consequence of my belief. So be it.

> It is the hallmark of the truth to be worth suffering for. In the deepest sense of the word, the evangelist must also be a martyr. If he is unwilling to do so, he should not lay his hand to the plow.[2]

The biblical passages at the beginning of this chapter are Jesus' command to spread the word of the truth of Christianity to all nations: the Great Commission. This book is my contribution to spreading the Word.

I have at various places in the book proposed various experiments to test my theory of Christianity. This is the way true science is done. Testable hypotheses are confronted by experiment. But the ultimate experiment will be development of the technology which will make use of the laws that permit matter to be converted into pure energy. This technology will solve our energy problems and incidentally provide a means to propel interstellar rockets. It will also enable us to create explosives of enormous power. As I have shown, this power must be given to us at some point in universal history, or unitarity, a fundamental law of quantum mechanics, would be violated.

Thus, we will be given this power, and human history in its usual sense will come to an end. But I have also shown that humanity will be brought back into existence as emulations in the computers of the far future. Each and every human who ever lived will be re-created, never to die again. So humanity will never die out ultimately, even if it ceases to play the major role in the universe that it now plays. The end-time of humanity as predicted by physics is so close to the end-time foretold by Christianity that I have proposed that the two are one and the same. Jesus will descend from the Second Hypostasis of the Cosmological Singularity, moving through the universes of the multiverse to take on human flesh once more. He will personally act to prevent the new superenergy source and the artificial intelligences from totally destroying humanity. He will instead guide both humanity and the new intelligences.

The Jews, God's once and always Chosen People, will, as predicted in the Revelation of John, convert in large numbers to Christianity and once again lead Christianity as they did in the first century. In the past twenty-five years, many Jews have come to play a larger positive role in the advance of Christianity than at any time since the first century. Jewish scientists have developed the physics essential for developing the theory I have described in this book. Jews have played a major role in increasing our understanding of the Turin Shroud. Before the Second Coming, I would expect to see a Jewish Pope.

Let us recall the main ideas developed in this book:

1. God is the Cosmological Singularity. A singularity is an entity that is outside of time and space—transcendent to space and time—and it is the only thing that exists that is not subject to the laws of physics. The Cosmological Singularity consists of three Hy-

postases: the Final Singularity, the All-Presents Singularity, and the Initial Singularity. These can be distinguished by using Cauchy sequences of different sorts of persons, so in the Cauchy completion, they become three distinct Persons. But still, the three Hypostases of the Singularity are just one Singularity. The Trinity, in other words, consists of three Persons but only one God.

2. Miracles never, never violate physical law. God—the Cosmological Singularity—created all that exists in the beginning of time knowing what He wanted to accomplish in universal history and adjusted the laws accordingly. He doesn't have to change the laws. Only an outsider deity would have to change the laws if he were to enter a universe created by another god. A "miracle" involving a violation of physical law is an idea associated with the Manichaean heresy, not traditional Christianity. A miracle violates only human ideas of how the laws of physics ought to operate, not the laws themselves. A miracle is a vastly improbable event from the human point of view but an event that would be seen to be inevitable if humans understood exactly where God intends universal history to go. Miracles are also direct manifestations of God's action in the universe, and this could be seen directly if we could measure the quantum coherence that goes from the event of the miracle directly into the Cosmological Singularity, which is God.

3. The greatest miracle is the Incarnation: the man Jesus is also God. I showed how this could in fact be true. We normal humans have analogues in only a limited number of universes in the multiverse, but Jesus' analogues could go all the way into the Second Hypostasis of the Cosmological Singularity. These persons collectively would *be* the Second Hypostasis, so the Second Hypostasis could equally well have been called the Son Singularity, or the Second Person of the Singularity, or the Second Person of the Trinity. This collectivity of persons is an example of a standard mathematical procedure called the Cauchy completion, so the difficult idea of how a man can also be God can be understood with a little advanced mathematics.

4. The Virgin Birth of Jesus is another miracle that even many scientists have difficulty accepting. They have believed that virgin birth never occurs in humans and that, even if it did, the virgin birth of a male would be impossible because, in a virgin birth, all the genetic information in the child would have to have come from

the mother, and a male necessarily has a Y chromosome, which exists only in males. But I pointed out that 1 male in 20,000 has two XX chromosomes and no Y chromosome, and the DNA on the Turin Shroud, purported to be the burial cloth of Jesus, has that signature.

5. The Resurrection of Jesus is a central miracle of Christianity in part because it shows that eventually we will all live again but also because Jesus' death and Resurrection were necessary in order to save the world. I showed how Jesus could have risen from the dead by making use of the baryon-annihilation process, which is responsible for all the matter that now exists in the cosmos. In the early universe, this process was used to convert radiation into matter. I proposed that Jesus reversed the process, converting the matter of his body into invisible radiation made up of neutrinos. An observer viewing this conversion of his body into neutrinos would see Jesus' body "dematerializing." Reversing the dematerialization would result in Jesus materializing apparently out of nothing. I suggested that Jesus used this process to give us a hint of how to do so ourselves. We must gain control of this process in order to prevent the violation of unitarity in the far future, a violation that would destroy the universe if it occurred. By dying and rising again, Jesus not only paid the price for our sins but also gave us the knowledge to save the entire universe from destruction. When God acts, He acts on many levels, with many levels of meaning, only some of which are obvious to humans at the time. I proposed several experiments to test my hypothesis that this matter-creating process was in operation 2,000 years ago. If my hypothesis is correct, we can use this process today. Developing the technology to do so is the ultimate test of the physics of Christianity.

It would make Christianity a branch of physics.

Appendix: Three Christian Creeds

The Apostles' Creed

Here is an English translation of the Apostles' Creed, from the Latin of about A.D. 700.

I believe in God the Father Almighty, Maker of heaven and earth.

And in Jesus Christ his only Son our Lord; who was conceived by the Holy Spirit, born of the Virgin Mary, suffered under Pontius Pilate, was crucified, dead, and buried; he descended into hell *[ad inferna];* the third day he rose again from the dead; he ascended into heaven, and sits on the right hand of God the Father Almighty; from thence he shall come to judge the quick and the dead.

I believe in the Holy Spirit; the holy catholic Church; the communion of saints; the forgiveness of sins; the resurrection of the body; and the life everlasting. AMEN.

The Athanasian Creed[1]

Whoever desires to be saved must above all things hold the Catholic faith. Unless a man keeps it in its entirety inviolate, he will assuredly perish eternally.

Now this is the Catholic faith, that we worship one God in Trinity and Trinity in unity, without either confusing the persons or dividing the substance. For the Father's person is one, the Son's another, the Holy Spirit's another; but the Godhead of the Father, the Son, and the Holy Spirit is one, their glory is equal, their majesty coeternal.

Such as the Father is, such is the Son, such also the Holy Spirit. The Father is uncreated, the Son uncreated, and Holy Spirit uncreated. The Father is infinite, the Son infinite, the Holy Spirit infinite. The Father is eternal, the Son eternal, the Holy Spirit eternal. Yet there are not three eternals, but one eternal; just as there are not three uncreateds or three infinites, but one uncreated and one infinite. In the same way the Father is omnipotent, the Son omnipotent, the Holy Spirit omnipotent; yet there are not three omnipotents, but one omnipotent.

Thus the Father is God, the Son God, the Holy Spirit God; and yet there are not three Gods, but One God. Thus the Father is Lord, the Son Lord, the Holy Spirit Lord; and yet there are not three Lords, but One Lord. Because just as we are obliged by Christian truth to acknowledge each person separately both God and Lord, so we are forbidden by the Catholic religion to speak of three Gods or Lords.

The Father is from none, not made nor created nor begotten. The Son is from the Father alone, not made nor created but begotten. The Holy Spirit is from the Father and the Son, not made nor created nor begotten but proceeding. So there is one Father, not three Fathers; one Son, not three Sons; one Holy Spirit, not three Holy Spirits. And in this Trinity there is nothing before or after, nothing greater or less, but all three persons are coeternal with each other and coequal. Thus in all things, as has been stated above, both Trinity in unity and unity in Trinity must be worshiped. So he who desires to be saved should think thus of the Trinity.

It is necessary, however, to eternal salvation that he should also faithfully believe in the Incarnation of our Lord Jesus Christ. Now the right faith is that we should believe and confess that our Lord Jesus Christ, the Son of God, is equally both God and man.

He is God from the Father's substance, begotten before time; and he is man from his mother's substance, born in time. Perfect God, perfect man composed of a rational soul and human flesh, equal to the Father in respect of his divinity, less than the Father in respect of his humanity.

Who, although he is God and man, is nevertheless not two but one Christ. He is one, however, not by the transformation of his divinity into flesh, but by the taking up of his humanity into God; one certainly not by confusion of substance, but by oneness of person. For just as rational soul and flesh are a single man, so God and man are a single Christ.

Who suffered for our salvation, descended to hell, rose from the dead, ascended to heaven, sat down at the Father's right hand, whence he will come to judge the living and the dead; at whose coming all men will rise again with their bodies, and will render an account of their deeds; and those who have behaved well go to eternal life, those who have behaved badly to eternal fire.

This is the Catholic faith. Unless a man believes it faithfully and steadfastly, he will not be able to be saved.

The Nicene Creed

Here is a literal translation of the Greek text of the Constantinopolitan form taken from *Catholic Encyclopedia,* the parentheses indicating the words altered or added in the Western liturgical form in present use in the Roman Catholic Church. I have added the original Greek in one place.

We believe (I believe) in one God, the Father Almighty, maker of heaven and earth, and of all things visible and invisible. And in one Lord Jesus Christ, the only begotten Son of God, and born of the Father before all ages. (God of God), light of light, true God of true God. Begotten not made, of the same substance *[homoousion]* as the Father, by whom all things were made. Who for us men and for our salvation came down from heaven. And was incarnate of the Holy Ghost and of the Virgin Mary and was made man; was crucified also for us under Pontius Pilate, suffered and was buried; and the third day rose again according to the Scriptures. And ascended into heaven, sits at the right hand of the Father, and shall come again with glory to judge the living and the dead, of whose Kingdom there shall be no end. And (I believe) in the Holy Ghost, the Lord and Giver of life, who proceeds from the Father (and the Son), who together with the Father and the Son is to be adored and glorified, who spoke by the Prophets. And (I believe in) one holy, catholic, and apostolic Church. We confess (I confess) one baptism for the remission of sins. And we look for (I look for) the resurrection of the dead and the life of the world to come. Amen.

Notes

1. Introduction

1 I have quoted the Fermi temperature for copper. See any graduate level textbook on solid-state physics or statistical thermodynamics, e.g., Reif 1965, p. 391. The only electrons that conduct electrical energy in the copper are those with energies above the Fermi temperature.

2. A Brief Outline of Modern Physics

1 Private remark by Stephen W. Hawking to Frank J. Tipler, September 1981, in Cambridge, England.
2 Gell-Mann 1992.
3 Weinberg 2001, p. 78.
4 Leggett 1986, p. 53.
5 Anderson 1986, p. 33.
6 Private conversation between Feynman and Kip S. Thorne, reported to Frank J. Tipler by Thorne.
7 Private remark by Leon Lederman to Frank J. Tipler.
8 Hamilton published the basic theory in two papers in 1834 and 1835, but he believed two equations were necessary. Jacobi pointed out in a 1938 paper that a single equation was sufficient. It is this that is now called the Hamilton-Jacobi equation. More details of the history can be found in Goldstine 1980, p. 176. The best introduction to the equation can be found in Landau and Lifshitz 1960, pp. 147–154.
9 Heisenberg 1930, pp. 66–76. The British physicist Nevill Mott (1929) ob-

tained the same answer at the same time. Both Heisenberg and Mott later won the Nobel Prize in Physics.

10 Everett 1957, reprinted in DeWitt and Graham 1973, pp. 141–149.

11 The Italian philosopher Giovanni Aguchhi made this Occam's razor argument against Copernicus in a letter to Galileo. See Drake 1978, p. 212.

12 The derivation is called Cox's Theorem. See Cox 1946, 1961 for more details. A more recent derivation of Cox's Theorem can be found in Jaynes 2003, pp. 17–33: the basic equations of Cox's Theorem are 2.63 and 2.64 on page 33 of Jaynes's book. Jaynes opines (2003, p. 686) that Cox's 1946 article "was the most important advance in the conceptual (as opposed to the simply mathematical) formulation of probability theory since Laplace." (Pierre-Simon, Marquis de Laplace, 1749–1827, was a great French mathematical physicist.) I agree.

13 This relabeling argument to get the numerical probabilities is called a *transformation group argument.* See Jaynes 2003, pp. 37–43, or Sivia 1996, pp. 106–110, for a more mathematically rigorous discussion.

14 Sivia 1996; Jaynes 2003.

15 This explanation of quantum tunneling, which is the correct one, generally cannot be found in textbooks on quantum mechanics written after 1950. The modern textbooks describe barrier penetration in terms of exact plane waves, which simplifies the calculation but obscures the physics, since plane waves are not allowable wave functions according to the postulates of quantum mechanics. (They are not square integrable and hence are not functions in a Hilbert space.) A correct explanation must use wave packets instead of plane waves, and this is the explanation I have given here. Older textbooks in quantum mechanics give the same explanation as I do; see, for example, Frenkel 1936, p. 73.

16 The energy-time uncertainty relation, which says $\Delta E \Delta t \geq h/4\pi$. The total energy is $E + \Delta E$, where E is the average energy, so we can increase the energy of a particle in a given universe as far above the average as we please, provided that we do it for a sufficiently short time, Δt.

17 For a proof of the unitarity of the S-matrix (a special case of the U-matrix discussed here; the S-matrix is a U-matrix with its initial time minus infinity and its final time plus infinity), see any book on relativistic quantum field theory, e.g., Bjorken and Drell 1965, pp. 145–146. Unitarity of the U-matrix is proved in a similar way.

18 See Misner, Thorne, and Wheeler 1973, chap. 12, and Tipler 1996b for discussions of Cartan's work on Newtonian gravitation.

3. Life and the Ultimate Future of the Universe

1 The usual number quoted for the age of the universe is 13.7 billion years. This age is obtained from three distinct observations: (1) measurements of fluctuations in the cosmic microwave background radiation, (2) measurements of the apparent luminosity of Type Ia supernovae, and (3) measurements of the Lyman alpha forest. I am personally dubious about number 3 because it yields a fluctuation spectrum inconsistent with numbers 1 and 2, while the fluctuation spectrum obtained by numbers 1 and 2 agrees with what one would expect assuming that the known physical laws are correct. Omitting number 3 gives 13.4 billion years for the age of the universe (and the multiverse).

2 This is somewhat analogous to Liouville's theorem in complex analysis, which says that all analytic functions other than constants have singularities either a finite distance from the origin of coordinates or at infinity.

3 Tipler 2005a.

4 A simple example of a universe with a final singularity but no event horizons is given by the metric $ds^2 = -dt^2 + R_0(t_s - t)^2[d\chi^2 + \sin^2\chi(d\theta^2 + \sin^2\theta d\phi^2)]$, where R_0 and t_s are constants. The constant R_0 measures the proper size of the universe, and t_s is the proper time at the final singularity. The metric is that of a 3-sphere universe, and $0 \leq \chi \leq \pi$, $0 \leq \theta \leq \pi$, $0 \leq \phi < 2\pi$. The coordinate χ is a radial coordinate for the 3-sphere, with $\chi = 0$ being the "north pole" and $\chi = \pi$ being the "south pole." If χ changes by 2π, the 3-sphere is circumnavigated. For a null geodesic (light ray path), θ and ϕ are constants and $ds^2 = 0$, so the variation of the radial distance is given by $\chi_f - \chi_i = \pm \int dt/[R_0(t_s - t)]$, where the upper and lower proper time limits of the integral are t_f and t_i, the final and initial times, respectively. The integral equals $(-1/R_0)ln\{(t_s - t_f)/(t_s - t_i)\}$, so as a null geodesic approaches the final singularity, its final time coordinate, t_f, approaches t_s, and thus $\chi_f - \chi_i$ approaches plus or minus infinity, the sign just denoting the direction of circumnavigation, which in turn means that each and every null geodesic circumnavigates the entire universe an infinite number of times before the singularity is reached, whatever the initial time, t_i, it starts its motion. Event horizons, therefore, cannot exist in this 3-sphere universe.

5 Matter-dominated or radiation-dominated universes are such 3-sphere universes. See Tipler 1996a, p. 435 (equation F.3), for the metric of a radiation-dominated universe in proper time and Tipler 1994a, p. 398 (equation B.2a), for the metric of a matter-dominated universe in conformal

time. Repeating the calculation of the preceding footnote, one sees that a light ray can travel only a fraction of the distance across the universe before it ends in the radiation case, and only one circumnavigation is possible in the matter case.

6 Feynman 1986, pp. 270–271.

7 For complete details, see Tipler 1994a, 2003, 2005a, 2005b, 2006.

8 Tipler 1994a, p. 435.

9 Hawking and Ellis 1973.

10 Guth 1981; Harrison 1970; Zel'dovich 1972.

11 Wilczek 2002; Quinn 2003, p. 35.

12 Bernstein 2001, pp. 22, 34. The critical mass of plutonium is 11 kg.

4. God as the Cosmological Singularity

1 Rudin 1964, p. 9.

2 Hawking and Ellis 1973.

3 Tipler 1994a, pp. 483–488.

4 Neumann 1955.

5 Bohr 1959, p. 209.

6 DeWitt and Graham 1973; Deutsch 1996.

7 Penrose 1989.

8 Tipler 1994a, pp. 259–265.

9 Ibid., sec. J, pp. 489–491.

10 Braaten and Clayton 1988, p. 12.

11 Pannenberg 1977, p. 122.

12 See Tipler 1994a, p. 313, and Pannenberg 1977, p. 126, for more discussion of this heresy.

13 Pannenberg 1977, pp. 120, 126, 160.

5. Miracles Do *Not* Violate Physical Law

1 Pannenberg 2002; see also Grant 1952, pp. 218–219.

2 Quoted in Swinburne 1970, p. 2.

3 Swinburne 1970.

4 Burns 1981.

5 Lewis 1978. The first edition of this book appeared in 1947.

6 Lewis 1970, p. 178. The essay quoted first appeared in the American journal *The Christian Century* 75 (November 26, 1958), pp. 1359–61, as stated in Lewis 1970, p. 15.

7 Lewis 1978, p. 59.

8 Ibid.

9 Straton 1924.

10 www.newadvent.org/cathen/10338a.htm. *Catholic Encyclopedia* was originally published in 1911 (New York: Appleton) but is now best accessed over the Internet.

11 E.g., Bjorken and Drell 1965, p. 145; Merzbacher 1970, pp. 500–502.

12 Hitchens 1995.

13 Quoted in ibid., pp. 25–26.

14 Quoted in ibid., pp. 26–27.

15 Quoted in ibid., pp. 61–62.

16 A photograph of the three children is in Brochado 1955, opposite the title page.

17 Jaki 1999, p. 11.

18 These photographs are reproduced in Brochado 1955, opposite p. 84.

19 Jaki 1999.

20 Ibid., p. 303.

21 Kottmeyer 2000.

22 Corliss 1984, p. 70.

23 Galileo 1953, p. 471.

24 Quoted in de Santillana 1959, p. 167.

25 Arberry 1996, p. 138. The passage is Arberry's translation of the Table Sura (chapter) of the Qur'an, the Sixth Sura in the traditional ordering.

26 Spencer 2002, p. 126.

27 Quoted in Hoodbhoy and Salam 1991, p. 54

28 Quoted in ibid., p. 105.

29 Nasr 1978, p. 230. This book is Nasr's Harvard University Ph.D. thesis. Nasr, who is now considered one of the world's leading experts on the relationship between science and Islam, started his undergraduate studies in physics at MIT, but he became disillusioned by the prevailing view of the time that science is merely instrumental and can never answer fundamental questions. He got his S.B. degree from MIT in physics but was more interested in the history of science courses given there, being particularly influenced by Giorgio de Santillana, the foremost Galileo scholar of his generation. Ten years after Nasr, I was a physics student at MIT and was also profoundly influenced by the history of science course I took from de Santillana. But I never lost my conviction that physics is the *only* source of answers to any question, in particular, the only source of answers to fundamental questions.

30 Hoodbhoy and Salam 1991, p. 29.
31 Hannam 2005.
32 See Tipler 2005a for the technical details.
33 Hoodbhoy and Salam 1991, p. 146.
34 www.zmag.org/content/ForeignPolicy/hoodbhoy0110.cfm.
35 Santillana 1961, pp. 280–313.
36 Cohen 1994, pp. 251–252.
37 Belloc 1938, pp. 82–96; Shafarevich 1980, pp. 18–79.
38 Shafarevich 1980, p. 19.
39 Trevor-Roper 1967, p. 92.
40 Ibid., n. 2.
41 Ibid., p. 92.
42 Ibid.
43 Ibid.
44 Ibid., p. 101.
45 Ibid., p. 117.
46 his "death to witches" passage is the most famous, but see also Deuteronomy 18:10, 2 Chronicles 33:6, Micah 5:12, and Nahum 3:4–5.
47 Trevor-Roper 1967, pp. 126–127.
48 Jaki 2000; Stark 2001.
49 Quoted in Aikman 2003, p. 5. Aikman is a former Beijing bureau chief for *Time* magazine.
50 See the first volume of Needham 1954, p. 581, or Cohen 1994, pp. 454–455, for a summary and extensive critical commentary.
51 Sorabji 1987, pp. 24–25, 52–53; Sambursky 1962, pp. 154–175.
52 Drake 1978.
53 Drake 1980.
54 De Santillana 1959 and Drake 1980.
55 Marsden 1994.
56 Quoted in Buckley 1997, pp. 29–30.
57 Ibid., pp. 28–31.
58 See Hicks 2004 for a summary of the nonsense now prevalent in the humanities faculties at most elite universities.
59 Darwin 1868, pp. 431–432. I used my personal copy of the first edition of Darwin's book.
60 Planck and Einstein 1932, p. 201.
61 It also had a negative influence on the theory of evolution. The population geneticist Richard Lewontin (1997) has emphasized that "the entire body of technical advance in experimental and theoretical evolutionary genetics of

the last fifty years has moved in the direction of emphasizing non-selective forces in evolution." Some of these, alas, are "random" forces.

62 Schönborn 2005.

63 Lewontin 1997.

64 Clarke 1717, pp. 15–17. Spelling and italics are Clarke's.

65 Caroline to Leibniz, January 10, 1716. Quoted in Alexander 1956, p. 193.

66 Note, however, that the Jewish New Year begins not on January 1, as in the standard calendar, but at the start of the lunar month of Tishrei, that is, on September 23 in 2006.

67 Hattaway et al. 2003, p. 103. The Chinese leaders are Brother Yun, Peter Xu Yongze (called the Billy Graham of China), and Enoch Wang.

68 Rutz 2005, p. 30.

69 Haddock 2001.

70 Spanos 1996.

71 For examples, see Rutz 2005, particularly pp. 59, 72–75, 185.

72 Ibid., p. 79.

73 Murray 1992.

6. The Christmas Miracle

1 Fesen, Hamilton, and Saken, 1989.

2 Brown 1993, pp. 165–201, 608–613.

3 Hughes 1976.

4 Sobel 1995, p. 168.

5 Neugebauer 1975, p. 8.

6 Evans 1998, pp. 17–18.

7 Dreyer 1963, p. 164 n.; Evans 1998, pp. 34–36.

8 Hughes 1976.

9 Ptolemy 1940, p. 143.

10 Ibid., p. 57.

11 Ibid., p. 259.

12 Ibid., p. 81.

13 Brown 1993, p. 549.

14 Ramsay 1915, pp. 295, 302.

15 Hughes 1976.

16 Ptolemy 1940, pp. 61, 197.

17 Hughes 1976.

18 Ibid.

19 Schaefer 1989.

20 Humphreys and Waddington 1983.
21 Espenak 1997.
22 Bruce 1981, p. 116; Theissen and Merz 1989, pp. 84–85.
23 Maier 1968, p. 13, n. 45; Theissen and Merz 1989, p. 85, n. 67.
24 Branch 1998; Galama et al. 1998.
25 Van den Bergh 2002; Fesen, Hamilton and Saken 1989.
26 Goldsmith 1989, pp. 16–17.
27 Stephenson and Green 2002; Hughes 1976.
28 Hughes 1976; Hughes 1979.
29 Edwards et al. 1977.
30 Hughes 1976.
31 Hughes 1979, p. 3.
32 Hughes 1976; Hughes 1979; Molnar 1999; Kidger 1999.
33 Hughes 1976.
34 Hattaway et al. 2003, p. 4.
35 Hughes 1976.
36 Noonan 1965.
37 Magnier et al. 1997.
38 De Vaucouleurs and Corwin 1985.
39 Fesen, Hamilton, and Saken 1989.
40 Baron 1998.
41 Galama et al. 1998.
42 Kulkarni et al. 1998.
43 The number of stars in the visible universe approximately equals $(4\pi/3)(1.88 \times 10^{\ 29} h^2 \text{ gm/cm}^3) \times (10 \text{ billion light-years})^3 \times [(\Omega_{sb})/(\text{mass of the Sun})] \approx 10^{20}$, where $h = 0.65$ is the Hubble factor, and Ω_{sb} is the fraction of mass in the form of baryons that are also in stars—roughly half of the universe's baryons are in stars, and the baryonic mass fraction is 0.04, so $\Omega_{sb} = 0.02$. The Hubble factor is a measure of the expansion of the universe. Hubble's constant H_0 is given in terms of the Hubble factor by $H_0 = h \times (100 \text{ km/sec/megaparsec})$.
44 The Sun's luminosity is 4×10^{33} ergs/sec, so the power output needed to equal 10^{20} stars is 4×10^{53} ergs/sec. The Newtonian gravitational potential energy of a constant density sphere of mass M and radius R is $-(3GM^2)/(5R)$, where G is the gravitational constant, so the change in gravitational potential energy from an initial radius, R_i, to a final radius, R_f, is therefore $\Delta PE = -GM^2[1/R_f - 1/R_i]$. The final radius cannot be smaller than the black hole radius $R_{BH} = 2GM/c^2 = 3 \text{ km } (M/M_{Sun})$, where M_{Sun} is the mass of the Sun, 2×10^{33} grams. Collapse from a radius that is signifi-

cantly larger than R_{BH} to R_{BH} will provide $3Mc^2/10$ of gravitational potential energy, which can be expressed in other forms, such as light and neutrinos. Thus, if a 1 solar mass stellar core collapses at light speed from 3,000 to 3 kilometers, the black hole radius, it would provide a power of 5×10^{53} ergs/sec. Ten solar masses collapsing at light speed to 30 kilometers would provide 5×10^{54} ergs/sec. The latter case is a more credible scenario for a hypernova power source, because it is unlikely that all the power released would appear as outward-moving radiation, and only in this case is the Newtonian approximation believable.

45 Hansen 1999.

46 Lai et al. 2001.

47 Brown 1993; Hughes 1976; Hughes 1979; Clark, Parkinson, and Stephenson 1977; Molnar 1999; Kidger 1999.

7. The Virgin Birth of Jesus

1 Johnson 1987, p. 90.

2 Ibid.

3 Ibid., p. 91.

4 Ibid.

5 Blenkinsopp 1964, p. 233.

6 Ibid.

7 Johnson 1987, pp. 90–91.

8 Kellner 1998, p. 119. I've not been able to find a hard copy of an English translation of Rashi's commentary on the Song of Songs, but an English translation does exist on a CD, having been made by the Judaica Press: *The Complete Tanach: Rashi.* This program is available for purchase online at http://www.hebrewlanguage.us/biblical.html.

9 Ibid., p. 24. Kellner translates Gersonides's commentary on Song of Songs 1:3. Kellner (p. 118) expresses amazement that Gersonides would translate *'alamot* as "virgins" given that Christians could use this translation to claim that it follows the same translation could be appropriate for Isaiah 7:14 (As I now indeed am claiming!).

10 Pannenberg 1977, pp. 141–150.

11 Ibid., p. 143.

12 Ibid., p. 120, n. 8, and p. 121.

13 Pannenberg, private communication to Frank J. Tipler.

14 See von Campenhausen 1964 and Boslooper 1962 for a detailed defense of this position.

15 Brown 1993; Laurentin 1986; Miguens 1980; Lewis 1978; Sayers 1978, p. 56.
16 Wright 2003, p. 596.
17 Bultmann 1960, p. 5.
18 Robinson 1976.
19 Stanton 2002.
20 Bowler 1971; Boylen 1984; Cole 1930; Preus 1970; Preus 1977; Stonehouse 1994.
21 Bultmann 1960, p. 11.
22 Harnack 1957, p. 30.
23 Zirkle 1936.
24 Jews censored the Talmud for self-protection in the sixteenth century. The anti-Christian comments were deleted, but the original version now exists in English translation. See, for example, Klinghoffer 2005 for a discussion of this passage on Mary's pregnancy.
25 See Beatty 1957 and Dawley and Bogart 1989 for an introduction to this enormous literature.
26 Murphy et al. 2000; Cassar et al. 1997; Cassar et al. 1998.
27 Groot et al. 2003.
28 Winston et al. 1991; Balakier et al. 1993; Levron et al. 1995; Marshall et al. 1998.
29 *Encyclopaedia Britannica,* "Multiple Births," 1967 and 2003 editions.
30 Becker et al. 1997.
31 Evett and Weir 1998, p. 116.
32 According to Bruce Weir (private communication). This formula is not in the literature. I therefore derived it.
33 Zhang 2000.
34 Marshall et al. 1998; see Rougier and Werb 2001 for a general discussion.
35 See Kuntziger and Bornens 2000 for a discussion of the peculiarity of the centromeres in primates.
36 The other two theories of how a male human can be born of a virgin are from Garza-Valdes 1999, p. 44, and Berry 1996, respectively.
37 Chapelle 1981; Guellean et al. 1984; Page et al. 1985; Andersson et al. 1986; Petit et al. 1987; Chapelle 1988.
38 Jegalian and Lahn 2001.
39 Diamond 2002, p. 704.
40 Casarino et al. 1995a.
41 Jegalian and Lahn 2001.
42 Zenteno et al. 1997; Abusheikha et al. 2001.
43 See Vermes 1973, p. 265; Vermes 2000, p. 225; Brown 1973; and espe-

cially Brown 1993, pp. 587–596, for a more extensive discussion of these inconsistencies.

44 Brown 1973; Brown 1993, p. 589.

45 Paragraphs 43 and 45, quoted by Vermes 1973, p. 265.

46 Brown 1993, pp. 81–84.

47 Ibid., p. 76.

48 Skaletsky et al. 2003.

49 Damon et al. 1989; Gove et al. 1997; Garza-Valdes 1999; Rogers 2005.

50 Rogers 2005.

51 Garza-Valdes 1999, pp. 49–53.

52 Adler 1996.

53 For example, see Wilson 1998; Wilson and Schwortz 2000.

54 See, for example, Wilson and Schwortz 2000, chap. 7, and pp. 151–156; see especially Wilson 1998, pp. 263–313.

55 Private communication to Frank J. Tipler, and see Garza-Valdes 1999, p. 49.

56 Damon et al. 1989.

57 Gove et al. 1997.

58 Garza-Valdes 1999, p. 137.

59 Ibid., pp. 115–119.

60 Berry 1996.

61 Garza-Valdes 1999, pp. 43–44.

62 Warner 1976, p. 35.

63 Garza-Valdes 1999.

64 Guscin 1998.

65 Casarino et al. 1995b.

66 Butler 2005, p. 114.

67 Ibid., pp. 564–566.

68 Margulis and Sagan 2002.

69 Margulis 1981; Margulis and Sagan 2002.

70 Clutton-Block 1999, pp. 36–37.

71 Pelikan 1996, p. 53.

72 Ibid., p. 102.

73 Coase 1988, Tipler 2007.

8. The Resurrection of Jesus

1 See Wright 2003; Habermas 1996; Pannenberg 1977; Pannenberg 2002 for examples.

2 Loftus and Ketcham 1991, p. 218.

3 A huge number of other examples of false memory can be found in Loftus 1991, Loftus and Ketcham 1996, 1997, 2003a, and 2003b.

4 Wilson and Schwortz 2000, p. 37.

5 Heller 1983, pp. 199–200; Jumper et al. 1984.

6 Pannenberg 1977, p. 98.

7 't Hooft 1976.

8 For examples, see Cheng and Li 1984; Rubakov and Shaposhnikov 1996; Weinberg 1996, chap. 23.

9 Weinberg 1996.

10 Heller 1983, p. 2.

11 Ibid., p. 201.

12 Wilson and Schwortz 2000, pp. 36–37.

13 Heller and Adler 1981.

14 Miller 1965.

15 Heller and Adler 1981.

16 Wilson and Schwortz 2000, p. 18.

17 Gove 1996, pp. 153–154.

18 Ibid., p. 261.

19 Halzen and Martin 1984, p. 273.

20 Heller 1983, p. 2.

21 See Kane 1993, app. C.

22 Collar 1996b argues that neutrinos have a much larger biological effect than is generally believed. I disagree, and use the standard estimate of the biological effect. See Cossairt and Marshall 1997 for a criticism of Collar.

23 I should also take into account the fact that, at very low energy, there is a coherence effect that substantially increases the elastic cross section between neutrinos and matter. See Drukier and Stodolsky 1984 for details. This effect is very important in understanding how supernovae blow off their outer envelopes. This effect would modify the details of the interaction of the neutrinos and the Shroud, but no essential conclusion would be changed. The math would be more complex, though, so I will omit it.

24 Wilson and Schwortz 2000, p. 49.

25 McMurry 1992.

26 Wilson and Schwortz 2000, pp. 37–39.

27 Scavone 1999, 2003. A summary of Scavone's thesis is given in Guscin 1998, p. 125, and in Wilson and Schwortz 2000, pp. 138–139, 169–173.

28 Wilson 1998; Wilson and Schwortz 2000.

29 Guscin 1998.

30 Weston 1904, 1913.
31 Chrétien de Troyes, 1999, p. 3, l. 67.
32 Braaten and Clayton 1988, p. 12.
33 Sayers 1978, p. 255.
34 Ratzinger 1998, pp. 58–59.
35 Fleischer, Price, and Walker 1975; and Fleischer 1998 are two standard references to nuclear track techniques in solids such as rocks.
36 Wilson 1998, pp. 105–107.
37 Fleischer 1998, table 4–1 on p. 89.
38 Ibid., p. 95.
39 Blake 1935, pp. 106–116.
40 Dawson 1889, p. 493.
41 Kenyon 1974, p. 226.
42 Ritmeyer and Ritmeyer 2004.
43 Kenyon and Moorey 1987, p. 178.
44 Murphy-O'Connor 1998, pp. 45–50.
45 Gil 1997, p. 373.
46 Since the Holy Sepulcher is inside the current Jerusalem city walls, there has been a claim since the nineteenth century that it is not the location of Jesus' tomb. Other locations outside the current city walls have been proposed. But the evidence is against these. See Ritmeyer and Ritmeyer 2004 and Murphy-O'Conner 1998 for the detailed archaeological argument. Finding nuclear particle tracks would decide the question conclusively.

9. The Grand Christian Miracle: The Incarnation

1 Deutsch 1986.
2 John Paul II, *Ecclesia de Eucharistia* (On the Eucharist in Its Relationship to the Church), par. 59.
3 Ibid., par. 55.
4 Ibid., par. 15.
5 Benedict XVI, Papal Homily, "The Sacrament of Unity," given May 29, 2005, at the Italian Eucharistic Congress.
6 I am grateful to Professor Wolfhart Pannenberg for an exchange of e-mails wherein he clarified for me the Lutheran doctrine of Real Presence and for these references to the Lutheran doctrinal position.

10. Anti-Semitism Is Anti-Christian

1 Johnson 1987, p. 127.

2 Luther 1955–1986, vol. 47, pp. 268–272. Quoted in Siemon-Netto 1995, p. 49.

3 Luther 1955–1986, vol. 45, pp. 200–201. Quoted by Siemon-Netto 1995, p. 49.

4 English translation available on the Internet at www.ncsj.org/AuxPages/032105RusPrav_trans.shtml.

5 Johnson 1987, p. 145.

6 Ibid.

7 Stark 1996.

8 Kurtzweil 2005.

9 Tipler 1994a, p. 23.

10 Barrow and Tipler 1986, p. 136.

11 Ibid.

12 Markoff 2003.

13 Copeland 2003.

14 Markoff 2005.

15 Rutz 2005, p. 41.

16 Stark 1992, 1996, 1999. According to Lester 2002 (p. 43, first column), Stark's view of religious movements is a "major force" in interpreting the future evolution of religion, at least among American sociologists of religion. It is interesting that many European sociologists disagree (Lester 2002, p. 44). Jenkins 2002.

17 Stark 1992, table on p. 16.

18 Stark 1996, p. 7.

19 Hattaway et al. 2003, p. 13. This figure appears in a number of sources, so it is probably fairly accurate.

20 Ibid., where the number is taken from *World Christian Encyclopedia.* Jenkins 2002 notes (p. 223, n. 3) that this number is double what other sources estimate. If the other sources are accurate, then Christians make up 3 percent rather than 7 percent of China's population. This would mean another decade or so will be required until China is majority Christian at the present rate of increase.

21 Jenkins 2002, p. 84.

22 Hattaway et al. 2003, p. 3.

23 Ibid.

11. The Problem of Evil and Free Will

1 Lay et al. 2005.

2 The main damage to New Orleans occurred not from the winds of the hurricane but from the flood that followed the levees being overtopped by the wind surge. The high point of the flood in my neighborhood was less than 1 meter from the beginning of my property line. A vastly improbable event, but as Chapter 4 makes clear, all direct acts of God are of exactly this nature.

3 Lovejoy 1960

4 Ponnuru 2006.

12. Conclusion

1 Sayers 1949, p. 28.

2 Joseph Cardinal Ratzinger, reply to Archbishop John May of St. Louis in meeting between Vatican officials and U.S. bishops, March 8, 1989. Quoted in Weigel 2002, p. 89.

Appendix

1 Taken from John Norman Davidson Kelly, *The Athanasian Creed* (New York: Harper & Row), 1964.

Bibliography

Abusheikha, N., et al. 2001. "XX Males Without SRY Gene and with Infertility." *Human Reproduction* 16 (4):717–18.

Adler, Alan D. 1996. "Updating Recent Studies on the Shroud of Turin." *Archaeological Chemistry, American Chemical Society Symposium Series* 625:223–28.

———. 1999. "The Nature of the Body Images on the Shroud of Turin." Chemistry Department of Western Connecticut University, preprint available on Web at www.shroud.com.

Aikman, David. 2003. *Jesus in Beijing: How Christianity Is Transforming China and Changing the Global Balance of Power.* Washington, D.C.: Regnery Publishing.

Alexander, H. G. 1956. *The Leibniz-Clarke Correspondence: With Extracts from Newton's "Principia" and "Optics."* Manchester: Manchester University Press.

Anderson, Philip. 1986. "Measurement in Quantum Theory and the Problem of Complex Systems." In *The Lesson of Quantum Theory: Niels Bohr Centenary Symposium,* ed. Jorrit de Boer, Erik Dal, and Ole Ulfbeck. Amsterdam: North Holland.

Andersson, Mea, et al. 1986. "Chromosome Y–Specific DNA Is Transferred to the Short Arm of the X Chromosome in Human XX Males." *Science* 233:786–88.

Arberry, Arthur John. 1996. *The Koran Interpreted.* New York: Simon & Schuster.

Balakier, H., et al. 1993. "Experimentally Induced Parthenogenetic Activation of Human Oocytes." *Human Reproduction* 8:740–43.

Baron, Eddie. 1998. "How Big Do Stellar Explosions Get?" *Nature* 395:635.

Barrow, John D., and F. J. Tipler. 1986. *The Anthropis Cosmological Principle.* Oxford, England: Oxford University Press.

Beatty, R. A. 1957. *Parthenogenesis and Polyploidy in Mammalian Development.* Cambridge: Cambridge University Press.

Becker, Albert, et al. 1997. "Twin Zygosity: Automated Determination with Microsatellites." *Journal of Reproductive Medicine* 42:260–66.

Belloc, Hilaire. 1938. *The Great Heresies.* Repr. Rockford, Ill.: Tan Books, 1991.

Bernstein, Jeremy. 2001. *Hitler's Uranium Club.* New York: Copernicus Books.

Berry, R. J. 1996. "The Virgin Birth of Christ." *Science and Christian Belief* 8:101–10.

Bjorken, James D., and Sidney D. Drell. 1965. *Relativistic Quantum Fields.* New York: McGraw-Hill.

Blake, George Stanfield. 1935. *The Stratigraphy of Palestine and Its Building Stones.* Jerusalem: Government Printing Office.

Blenkinsopp, Joseph. 1964. *Isaiah 1–39.* Vol. 19 of *The Anchor Bible.* New York: Doubleday.

Blum, Howard. 2003. *The Eve of Destruction: The Untold Story of the Yom Kippur War.* New York: HarperCollins.

Bohr, Niels. 1959. "Discussion with Einstein on Epistemological Problems in Atomic Physics." In *Albert Einstein: Philosopher-Scientist,* ed. P. A. Schilpp. New York: Harper & Row.

Boslooper, Thomas. 1962. *The Virgin Birth.* Philadelphia: Westminster Press.

Bowler, Peter J. 1971. "Preformation and Pre-existence in the Seventeenth Century: A Brief Analysis." *Journal of the History of Biology* 4 (2): 221–44.

Boylen, Michael. 1984. "The Galenic and Hippocratic Challenges to Aristotle's Conception Theory." *Journal of the History of Biology* 17 (1): 83–112.

Braaten, Carl E., and Philip Clayton. 1988. *The Theology of Wolfhart Pannenberg.* Minneapolis: Augsburg Press.

Branch, D. 1998. "Type Ia Supernovae and the Hubble Constant." *Annual Review of Astronomy and Astrophysics* 36:17–55.

Brochado, Costa. 1955. *Fatima in the Light of History.* Milwaukee: Bruce Publishing.

Brown, Raymond E. 1973. *The Virginal Conception and Bodily Resurrection of Jesus.* New York: Paulist Press.

Brown, Raymond E. 1993. *The Birth of the Messiah,* updated ed. New York: Doubleday.

Bruce, F. F. 1981. *The New Testament Documents: Are They Reliable?* 6th ed. Grand Rapids, Mich.: Eerdmans.

Buckley, William F., Jr. 1997. *Nearer, My God: An Autobiography of Faith.* New York: Harcourt, Brace.

Bultmann, Rudolf. 1960. "New Testament and Mythology." In *Kerygma and Myth: A Theological Debate,* ed. Hans Werner Bartsch. London: SPCK Press.

Burns, R. M. 1981. *The Great Debate on Miracles: From Joseph Granville to David Hume.* Lewisburg, Pa.: Bucknell University Press.

Butler, John M. 2005. *Forensic DNA Typing: Biology, Technology, and Genetics of STR Markers,* 2nd ed. New York: Elsevier Academic Press.

Campenhausen, Hans von. 1964. *The Virgin Birth in the Theology of the Ancient Church.* Naperville, Ill.: Allenson Press.

Casarino, Lucia, et al. 1995a. "HLA-DQA1 and Amelogenin Coamplification: A Handy Tool for Identification." *Journal of Forensic Sciences* 3:456–58.

Casarino, Lucia, et al. 1995b. "Ricerca dei Polimorfismi del DNA Sulla Sindone e Sul Sudario di Oviedo" (in Italian; English translation: "Research on the Polymorphisms of the DNA on the Turin Shroud and the Oviedo Cloth"), *Sindon N.S. Quad.* 8 (December): 36–47.

Cassar, G., et al. 1997. "Observations on Ploidy of Cells and on Reproductive Performance in Parthenogenetic Turkeys." *Poultry Science* 77:1457–62.

Cassar, G., et al. 1998. "Differentiating Between Parthenogenetic and Positive Development Embryos in Turkeys by Molecular Sexing." *Poultry Science* 77:1463–68.

Chapelle, Albert de la. 1981. "The Etiology of Maleness in XX Men." *Human Genetics* 58:105–116.

Chapelle, Albert de, et al. 1988. "Invited Editorial: The Complicated Issue of Human Sex Determination." *American Journal of Human Genetics* 43:1–3.

Cheng, Ta-Pei, and Ling-Fong Li. 1984. *Gauge Theory of Elementary Particle Physics.* Oxford: Oxford University Press.

Chrétien de Troyes. 1999. *Perceval: The Story of the Grail,* trans. Burton Raffel. New Haven: Yale University Press.

Clark, David H., John H. Parkinson, and F. Richard Stephenson. 1977. "An Astronomical Re-Appraisal of the Star of Bethlehem: A Nova in 5 B.C." *Quarterly Journal of the Royal Astronomical Society* 18:443–49.

Clarke, Samuel. 1717. *A Collection of Papers Which Passed Between the Late Learned Mr. Leibnitz, and Dr. Clarke, in the Years 1715 and 1716, Relating to the Principles of Natural Philosophy and Religion.* London: James Knapton.

Clutton-Block, Juliet. 1999. *A Natural History of Domesticated Mammals,* 2nd ed. Cambridge: Cambridge University Press.

Coase, Ronald H. 1988. *The Firm, the Market, and the Law.* Chicago: University of Chicago Press.

Cohen, H. Floris. 1994. *The Scientific Revolution: A Historiographical Inquiry.* Chicago: University of Chicago Press.

Cole, F. J. 1930. *Early Theories of Sexual Generation.* Oxford: Clarendon Press.

Collar, Juan I. 1996a. "Comment on 'Limits on Dark Matter Using Ancient Mica.' " *Physical Review Letters* 76:331.

Collar, Juan I. 1996b. "Biological Effects of Stellar Collapse Neutrinos." *Physical Review Letters* 76:999–1002 and 78:1395.

Collar, Juan I., and F. T. Avignone III. 1995. "Nuclear Tracks from Cold Dark Matter Interactions in Mineral Crystals: A Computational Study." *Nuclear Instruments and Methods in Physics Research* B95:349–54.

Copeland, Michael V. 2003. "Cray Inc.'s Revenge." *Business 2.0* (September).

Corliss, William R. 1984. *Rare Halos, Mirages, Anomalous Rainbows, and Related Electro-magnetic Phenomena: A Catalog of Geophysical Anomalies.* Glen Arms, Md.: Sourcebook Project.

Cossairt, J. D., and E. T. Marshall. 1997. "Comment on 'Biological Effects of Stellar Collapse Neutrinos.' " *Physical Review Letters* 78:1394.

Cox, Richard T. 1946. "Probability, Frequency, and Reasonable Expectation." *American Journal of Physics* 14:1–13.

Cox, Richard T. 1961. *The Algebra of Probable Inference.* Baltimore: Johns Hopkins University Press.

Crabtree, Adam. 1985. *Multiple Man: Explorations in Possession and Multiple Personality.* New York: Praeger Scientific Books.

Damon, P. E., et al. 1989. "Radiocarbon Dating of the Shroud of Turin." *Nature* 337:611–15.

Darwin, Charles. 1868. *The Variation of Animals and Plants Under Domestication,* vol. 2. London: John Murray.

Dawley, Robert M., and James P. Bogart, eds. 1989. *Evolution and Ecology of Unisexual Vertebrates.* Museum Bulletin 466. Albany: New York State Museum Press.

Dawson, Sir John W. 1889. *Modern Science in Bible Lands.* New York: Harper.

Deutsch, David. 1986. "Three Connections Between Everett's Interpretation and Experiment." In *Quantum Concepts in Space and Time,* ed. R. Penrose and C. J. Isham. Oxford, England: Oxford University Press.

Deutsch, David. 1996. *The Fabric of Reality.* London: Basic Books.

De Vaucouleurs, G., and H. G. Corwin. 1985. "S Andromedae 1885: A Centennial Review." *Astrophysical Journal* 295:287–304.

DeWitt, Bryce, and Niel Graham. 1973. *The Many-Worlds Interpretation of Quantum Mechanics.* Princeton: Princeton University Press.

Diamond, Jared. 2002. "Evolution, Consequences and Future of Plant and Animal Domestication." *Nature* 418:700–707.

Drake, Stillman. 1978. *Galileo at Work.* Chicago: University of Chicago Press.

Drake, Stillman. 1980. *Galileo: Oxford Past Masters Series.* New York: Hill & Wang.

Dreyer, J. L. E. 1963. *Tycho Brahe.* New York: Dover.

Driscoll, John T. 2003. "Miracle." *Catholic Encyclopedia.* www.newadvent .org/cathen.

Drukier, A., and L. Stodolsky. 1984. "Principles and Applications of a Neutral-Current Detector for Neutrino Physics and Astronomy." *Physical Review D* 30:2295–2309.

Durrani, S. A., and R. K. Bull. 1987. *Solid State Nuclear Track Detection: Principles, Methods, and Applications.* New York: Pergamon Press.

Edwards, Ormand, et al. 1977. "The Star of Bethlehem." *Nature* 268:565–67.

Espenak, F. 1997. "Historical Solar Eclipses." http://sunearth.gsfc.nasa.gov/eclipse/SEhistory/SEhistory.html.

Evans, J. 1998. *The History and Practice of Ancient Astronomy.* New York: Oxford University Press.

Everett, Hugh. 1957. "Relative State Formulation of Quantum Mechanics." *Reviews of Modern Physics* 29: 454–459.

Evett, Ian W., and Bruce S. Weir. 1998. *Interpreting DNA Evidence: Statistical Genetics for Forensic Scientists.* Sunderland, Mass.: Sinauer Associates.

Fanti, Giulio, and Roberto Maggiolio. 2004. "The Double Superficiality of the Frontal Image of the Turin Shroud." *Journal of Optics A: Pure and Applied Optics* 6:491–503.

Fesen, Robert A., Andrew J. S. Hamilton, and Jon M. Saken. 1989. "Discovery of the Remnant of S Andromadae (SN 1885) in M31." *Astrophysical Journal* 341:L55–L57.

Feynman, Richard P. 1986. *Surely You're Joking, Mr. Feynman.* New York: Bantam Books.

Fleischer, Robert L. 1998. *Tracks to Innovation: Nuclear Tracks in Science and Technology.* Berlin: Springer-Verlag.

Fleischer, Robert L., P. Buford Price, and Robert M. Walker. 1975. *Nuclear Tracks in Solids.* Berkeley: University of California Press.

Frenkel, J. 1936. *Wave Mechanics,* 2nd ed. Oxford: Oxford University Press. Repr., New York: Dover Publications, 1950.

Friedgut, Theodore. 1994. *Antisemitism and Its Opponents in the Russian Press: From Perestroika to the Present.* ACTA 3. Jerusalem: Hebrew University Press.

Friesen, James G. 1991. *Uncovering the Mystery of MPD.* San Bernardino, Calif.: Here's Life Publishers.

Galama, T. J., et al. 1998. "An Unusual Supernova in the Error Box of the Gamma-Ray Burst of 25 April 1998." *Nature* 395:670–72.

Galileo Galilei. 1953. *Dialogue on the Great World Systems,* trans. Giorgio de Santillana. Chicago: University of Chicago Press.

Garza-Valdes, Leoncio A. 1999. *The DNA of God?* New York: Doubleday.

Gell-Mann, Murray. 1992. "[Remarks on Bryce DeWitt's Paper] 'DeCoherence Without Complexity and Without an Arrow of Time.' " In *Physical Origins of Time Asymmetry,* ed. J. J. Halliwell, J. Perez-Marcader, and W. H. Zurek. Cambridge: Cambridge University Press.

George, Robert P. 2001. *A Clash of Orthodoxies: Law, Religion, and Morality in Crisis.* Wilmington, Del.: ISI Books.

Gil, Moshe. 1997. *A History of Palestine,* 634–1099. Cambridge: Cambridge University Press.

Goldsmith, D. 1989. *Supernova! The Exploding Star of 1987.* New York: St. Martin's Press.

Goldstine, Herman H. 1980. *A History of the Calculus of Variations.* Berlin: Springer-Verlag.

Gove, Harry E. 1996. *Relic, Icon, or Hoax? Carbon Dating the Shroud of Turin.* Bristol: Institute of Physics Press.

Gove, Harry E., et al. 1997. "A Problematic Source of Organic Contamination in Linen." *Nuclear Instruments and Methods in Physics Research B* 123:504–7.

Grant, Robert M. 1952. *Miracle and Natural Law in Graeco-Roman and Early Christian Thought.* Amsterdam: North Holland.

Groot, T. V. M., et al. 2003. "Molecular Genetic Evidence for Parthenogenesis in the Burmese Python, *Python molurus bivittatus.*" *Heredity* 90:130–35.

Guellean, Georges, et al. 1984. "Human XX Males with Y Single-Copy DNA Fragments." *Nature* 307:172–73.

Guscin, Mark. 1998. *The Oviedo Cloth.* Cambridge: Lutterworth Press.

Guth, Alan. 1981. "The Inflationary Universe: A Possible Solution to the Horizon and Flatness Problems." *Physical Review* D23:347–56.

Habermas, Gary R. 1989. "Resurrection Claims in Non-Christian Religions." *Religious Studies* 25:167–77.

Habermas, Gary R. 1996. *The Historical Jesus: Ancient Evidence for the Life of Christ.* Joplin, Mo.: College Press.

Habermas, Gary R. 2001a. "The Late Twentieth-Century Resurgence of Naturalistic Responses to Jesus' Resurrection." *Trinity Journal* 22:179–96.

Habermas, Gary R. 2001b. "Explaining Away Jesus' Resurrection: The Recent Revival of Hallucination Theories." *Christian Research Journal* 23 (4):26–31, 47–50.

Haddock, Deborah Bray. 2001. *The Dissociative Identity Disorder Sourcebook.* Chicago: Contemporary Books.

Halzen, Francis, and Alan D. Martin. 1984. *Quarks and Leptons.* New York: Wiley.

Hannam, James. 2005. "The Foundation and Loss of the Royal and Serapeum Libraries of Alexandria." http://www.bede.org.uk/Library2.htm.

Hansen, Brad M. S. 1999. "On the Frequency and Remnants of Hypernovae." *Astrophysical Journal* 512: L117–L120.

Harnack, Adolf. 1957. *What Is Christianity?* New York: Harper Torchbooks.

Harrison, Edward. 1970. "Fluctuations at the Threshold of Classical Cosmology." *Physical Review* D1:2726–30.

Hattaway, Paul, and Brother Yun. 2002. *The Heavenly Man: The Remarkable True Story of Chinese Christian Brother Yun.* London: Monarch Books.

Hattaway, Paul, et al. 2003. *Back to Jerusalem: Three Chinese House Church Leaders Share Their Vision to Complete the Great Commission.* Waynesboro, Ga.: Gabriel Publishing.

Hawking, Stephen W., and George F. R. Ellis. 1973. *The Large-Scale Structure of Space-Time.* Cambridge: Cambridge University Press.

Heisenberg, Werner. 1930. *The Physical Principles of the Quantum Theory.* Chicago: University of Chicago Press.

Heller, John H. 1983. *Report on the Shroud of Turin.* Boston: Houghton Mifflin.

Heller, John H., and Alan D. Alder. 1981. "A Chemical Investigation of the Shroud of Turin." *Canadian Society of Forensic Science Journal* 14:81–103.

Hicks, Stephen R. C. 2004. *Explaining Postmodernism.* Milwaukee: Scholarly Publishing.

Hitchens, Christopher. 1995. *The Missionary Position: Mother Teresa in Theory and Practice.* New York: Verso.

Hoodbhoy, Pervez, and Mohammed Abdus Salam. 1991. *Islam and Science: Religious Orthodoxy and the Battle for Rationality.* London: Zed Books.

Hughes, D. W. 1976. "The Star of Bethlehem." *Nature* 264:513–17.

Hughes, D. W. 1979. *The Star of Bethlehem.* New York: Pocket Books.

Humphreys, Colin J., and W. G. Waddington. 1983. "Dating the Crucifixion." *Nature* 306:743–46.

Iqbal, Muzaffar. 2002. *Islam and Science.* Burlington, Vt.: Ashgate Publishing.

Jackson, John P. 1990. "Is the Image on the Shroud Due to a Process Heretofore Unknown to Modern Science?" *Shroud Spectrum International* 34:11–20.

Jaki, Stanley L. 1999. *God and the Sun at Fatima.* Pinckney, Mich.: Real View Books.

Jaki, Stanley L. 2000. *The Savior of Science.* Grand Rapids, Mich.: Eerdmans.

Jaynes, Edward T. 1985. "Bayesian Methods: General Background." In *Maximum Entropy and Bayesian Methods in Applied Statistics,* ed. J. H. Justice, 1–25. Cambridge: Cambridge University Press.

Jaynes, Edward T. 2003. *Probability Theory: The Logic of Science.* Cambridge: Cambridge University Press.

Jeffreys, Sir Harold. 1939. *Theory of Probability.* Oxford: Clarendon Press.

Jegalian, Karin, and Bruce T. Lahn. 2001. "Why the Y Is So Weird." *Scientific American,* February, 56–61.

Jenkins, Philip. 2002. *The Next Christendom: The Coming of Global Christianity.* New York: Oxford University Press.

Johnson, Paul. 1987. *A History of the Jews.* London: Weidenfeld & Nicolson.

Jumper, Eric J., et al. 1984. "A Comprehensive Examination of the Various Stains and Images on the Shroud of Turin." *Archaeological Chemistry III. Advances in Chemistry Series* (American Chemical Society) 205:447–76.

Kane, Gordon. 1993. *Modern Elementary Particle Physics,* undated ed. Reading, Mass.: Perseus Books.

Kellner, Menachem. 1998. *Commentary on the Song of Songs by Levi ben Gershom (Gersonides).* New Haven: Yale University Press.

Kenyon, Kathleen M. 1974. *Digging Up Jerusalem.* London: Ernest Benn.

Kenyon, Kathleen M., and P. R. S. Moorey. 1987. *The Bible and Recent Archaeology.* London: British Museum.

Kidger, M. 1999. *The Star of Bethlehem.* Princeton: Princeton University Press.

Klinghoffer, David. 2005. *Why the Jews Rejected Jesus.* New York: Doubleday.

Kottmeyer, Martin. 1994. "The Eyes That Spoke." *Skeptical Briefs Newsletter,* September. www.csicop.org/sb/9409/eyesthat.html.

Kottmeyer, M. T. 2000. "Review of *God and the Sun at Fatima,* by Stanley L. Jaki," *Magonia Supplement* 29. www.magonia.demon.co.uk/arc/00/ms29.html.

Kulkarni, S. R., et al. 1998. "Radio Emission from the Unusual Supernova 1998bw and Its Association with the γ-burst of 25 April 1998." *Nature* 395:663–69.

Kuntziger, T., and M. Bornens. 2000. "The Centrosome and Parthenogenesis." *Current Topics in Developmental Biology* 49:1–25.

Kurtzweil, Ray. 2005. *The Singularity Is Near: When Humans Transcend Biology.* New York: Viking Books.

Lai, S.-P., et al. 2001. "A Critical Examination of Hypernova Remnant Candidates in M101.I.MF83." *Astrophysical Journal* 547:754–64.

Landau, Lev D., and E. M. Lifshitz. 1960. *Mechanics.* Reading, Mass.: Addison-Wesley.

Laurentin, Rene. 1986. *The Truth of Christmas Beyond the Myths: The Gospels of the Infancy of Christ.* Petersham, Mass.: St. Bede's Publications.

Lay, Thorne, et al. 2005. "The Great Sumatra-Andaman Earthquake of 26 December 2004." *Science* 308 (5725): 1127–33.

Leggett, Anthony J. 1986. "Quantum Mechanics at the Macroscopic Level." In *The Lesson of Quantum Theory: Niels Bohr Centenary Symposium,* ed. Jorrit de Boer, Erik Dal, and Ole Ulfbeck. Amsterdam: North Holland.

Lester, Toby. 2002. "Oh, Gods!" *Atlantic Monthly* 289 (2): 37–45.

Levron, J., et al. 1995. "Highly Effective Method of Human Oocyte Activation." *Zygote* 3:157–61.

Lewis, Clive S. 1970. *God in the Dock.* Grand Rapids, Mich.: Eerdmans.

Lewis, Clive S. 1978. *Miracles.* New York: Macmillan Paperbacks.

Lewontin, Richard. 1997. "Billions and Billions of Demons." *New York Times Book Review,* January 9.

Loftus, Elizabeth F. 1996. *Eyewitness Testimony.* Cambridge, UK: Cambridge University Press.

Loftus, Elizabeth F. 1997. "Creating False Memories." *Scientific American* 277 (3): 70–75.

Loftus, Elizabeth F. 2003a. "Our Changeable Memories: Legal and Practical Implications." *Nature Reviews: Neuroscience* 4:231–34.

Loftus, Elizabeth F. 2003b. "Make-Believe Memories." *American Psychologist* 58:864–73.

Loftus, Elizabeth F., and Katherine Ketcham. 1991. *Witness for the Defense.* New York: St. Martin's Press.

Loftus, Elizabeth F., and Katherine Ketcham. 1994. *The Myth of Repressed Memory.* New York: St. Martin's Press.

Lopez, M., et al. 1995. "SRY Alone Can Induce Normal-Male Sexual-Differentiation." *American Journal of Medical Genetics* 55 (3): 356–58.

Lovejoy, Arthur O. 1960. *The Great Chain of Being: A Study of the History of an Idea.* New York: Harper.

Luther, Martin. 1955–1986. *Luther's Works, in 55 Volumes,* ed. Jaroslav Pelikan and Helmut T. Lehmann. St. Louis: Concordia Publishing House and Philadelphia: Fortress Press.

McMurry, John. 1992. *Organic Chemistry,* 3rd ed. Belmont, Calif.: Wadsworth.

Magnier, E. A., F. A. Primini, and S. Prins. 1997. "ROSAT HRI Observations of M31 Supernova Remnants." *Astrophysical Journal* 490:649–52.

Maier, Paul L. 1968. "Sejanus, Pilate, and the Date of the Crucifixion." *Church History* 37:3–13.

Mannucci, A., et al. 1994. "Forensic Application of a Rapid and Quantitative DNA Sex Test by Amplification of the X-Y Homologous Gene Amelogenin." *International Journal of Legal Medicine* 106:190–93.

Margulis, Lynn. 1981. *Symbiosis in Cell Evolution.* San Francisco: Freeman.

Margulis, Lynn, and Dorion Sagan. 2002. *Acquiring Genomes.* New York: Basic Books.

Markoff, Mark. 2003. "Low-Cost Supercomputer Put Together from 1,100 PC's." *New York Times,* October 22.

Markoff, Mark. 2005. "A New Arms Race to Build the World's Mightiest Computer." *New York Times,* August 19.

Marsden, George. 1994. *The Soul of the American University: From Protestant Establishment to Established Unbelief.* New York: Oxford University Press.

Marshall, Vivienne S., et al. 1998. "Parthenogenetic Activation of Marmoset (*Callithrix jacchus*) Oocytes and the Development of Marmoset Parthenogenomes in Vitro and in Vivo." *Biology of Reproduction* 59:1491–97.

Mazzoni, Giuliana A. L., Elizabeth F. Loftus, and Irving Kirsch. 2001. "Changing Beliefs About Implausible Autobiographical Events: A Little Plausibility Goes a Long Way." *Journal of Experimental Psychology: Applied* 7:51–59.

Merzbacher, Eugen. 1970. *Quantum Mechanics,* 2nd ed. New York: Wiley.

Miguens, Manuel. 1980. "The Infancy Narratives and Critical Biblical Method." *Communio* 7:24–54.

Miller, Charles E. 1965. "Hydrogenation with Diimide." *Journal of Chemical Education* 42:254–59.

Misner, Charles W., Kip S. Thorne, and John A. Wheeler. 1973. *Gravitation.* San Francisco: Freeman.

Molnar, M. R. 1999. *The Star of Bethlehem.* New Brunswick, N.J.: Rutgers University Press.

Mott, Neville F. 1929. "The Wave Mechanics of α-Ray Tracks." *Proceedings of the Royal Society of London A* 126:79–84.

Murphy, Robert W., et al. 2000. "A Fine Line Between Sex and Unisexuality:

The Phylogenetic Constraints on Parthenogenesis in Lacertid Lizards." *Zoological Journal of the Linnaean Society* 130:527–49.

Murphy-O'Connor, Jerome. 1998. *The Holy Land: An Oxford Archaeological Guide from Earliest Times to 1700.* Oxford: Oxford University Press.

Murray, William J. 1992. *My Life Without God.* Eugene, Ore.: Harvest House Publishers.

Nasr, Seyyed Hossein. 1978. *An Introduction to Islamic Cosmological Doctrines.* Boulder, Colo.: Shambhala Books.

Needham, Joseph. 1954. *Science and Civilization in China,* vol. 1. Cambridge: Cambridge University Press.

Neugebauer, Otto. 1975. *A History of Ancient Mathematical Astronomy, Part 1.* Berlin: Springer.

Neumann, John von. 1955. *The Mathematical Foundations of Quantum Mechanics.* Princeton: Princeton University Press.

Noonan, John T. 1965. *Contraception: A History of Its Treatment by the Catholic Theologians and Canonists.* Cambridge, Mass.: Harvard University Press.

Page, David C., et al. 1985. "Chromosome-Specific DNA in Related Human XX Males." *Nature* 315:224–26.

Pannenberg, Wolfhart. 1977. *Jesus: God and Man,* 2nd ed. Philadelphia: Westminster Press.

Pannenberg, Wolfhart. 1994. *Systematic Theology,* vol. 2. Grand Rapids, Mich.: Eerdmans.

Pannenberg, Wolfhart, 2002. "The Concept of Miracle." *Zygon* 37:759–62.

Pearson, Brook W. R. 1999. "The Lucan Censuses Revisited." *Catholic Biblical Quarterly* 61:262–282.

Pelikan, Jaroslav. 1996. *Mary Through the Centuries: Her Place in the History of Culture.* New Haven: Yale University Press.

Penrose, Roger. 1989. *The Emperor's New Mind.* Oxford: Oxford University Press.

Petit, Christine, et al. 1987. "An Abnormal Terminal X-Y Interchange Accounts for Most but Not All Cases of Human XX Maleness." *Cell* 49:595–602.

Planck, Max, and Albert Einstein. 1932. *Where Is Science Going?* New York: W. W. Norton.

Podlecki, Anthony J. 1989. *Aeschylus: Eumenides.* Warminster: Aris & Phillips.

Ponnuru, Ramesh. 2006. *The Party of Death.* Washington, D.C.: Regnery.

Preus, Anthony. 1970. "Science and Philosophy in Aristotle's *Generation of Animals.*" *Journal of the History of Biology* 3 (1): 1–52.

Preus, Anthony. 1977. "Galan's Criticism of Aristotle's Conception Theory." *Journal of the History of Biology* 10 (1): 65–85.

Ptolemy, Claudius. 1940. *Tetrabiblos,* trans. F. E. Robbins. Cambridge, Mass.: Harvard University Press.

Quinn, Helen R. 2003. "The Asymmetry Between Matter and Antimatter." *Physics Today* 56 (2): 30–35.

Ramsay, Sir William M. 1898. *Was Christ Born in Bethlehem?* London: Hodder & Stoughton.

Ramsay, Sir William M. 1912. "Luke's Narrative of the Birth of Christ." *Expository Times* 4:385–407.

Ramsay, Sir William M. 1915. *The Bearing of Recent Discovery on the Trustworthiness of the New Testament.* London: Hodder & Stoughton.

Ratzinger, Joseph. 1998. *Milestones: Memoirs 1927–1977.* San Francisco: Ignatius Press.

Reif, Frederick. 1965. *Fundamentals of Statistical and Thermal Physics.* New York: McGraw-Hill.

Ritmeyer, Leen, and Kathleen Ritmeyer. 2004. *Jerusalem in the Year 30 A.D.* Jerusalem: Carta.

Robinson, John A. T. 1976. *Redating the New Testament.* London: SCM Press.

Rogers, Raymond N. 2005. "Studies on the Radiocarbon Sample from the Shroud of Turin." *Thermochimica Acta* 425:189–94.

Rougier, Nathalie, and Zena Werb. 2001. "Minireview: Parthenogenesis in Mammals." *Molecular Reproduction and Development* 59:468–74.

Rubakov, V. A., and Mikhail E. Shaposhnikov. 1996. "Electroweak Baryon Number Non-Conservation in the Early Universe and in High-Energy Collisions." *Physics-Uspekhi* 39 (5): 461–502.

Rudin, Walter. 1964. *Principles of Mathematical Analysis.* New York: McGraw-Hill.

Rutz, James. 2005. *Megashift.* Colorado Springs: Empowerment Press.

Sambursky, Samuel. 1962. *The Physical World of Late Antiquity.* New York: Basic Books.

De Santillana, Giorgio. 1959. *The Crime of Galileo.* Chicago: University of Chicago Press.

De Santillana, Giorgio. 1961. *The Origins of Scientific Thought.* New York: Mentor Books.

Sayers, Dorothy L. 1949. *Creed or Chaos?* New York: Harcourt, Brace.

Sayers, Dorothy L. 1978. *The Whimsical Christian.* New York: Macmillan.

Scavone, Daniel. 1999. "Joseph of Arimathea, the Holy Grail, and the Edessa Icon." *Arthuriana* 9 (4):2–31.

Scavone, Daniel. 2003. "British King Lucius, the Grail, and Joseph of Arimathea: The Question of Byzantine Origins." *Publications of the Medieval Association of America* 10: 101–142.

Schaefer, Bradley E. 1989. "Dating the Crucifixion." *Sky and Telescope* 77 (April): 374.

Schönborn, Christoph. 2005. "Finding Design in Nature." *New York Times,* July 7.

Shafarevich, Igor R. 1980. *The Socialist Phenomenon.* New York: Harper & Row.

Siemon-Netto, Uwe. 1995. *The Fabricated Luther: The Rise and Fall of the Shirer Myth.* St. Louis: Concordia Publishing House.

Sivia, D. S. 1996. *Data Analysis: A Bayesian Tutorial.* Oxford: Clarendon Press.

Skaletsky, Helen, et al. 2003. "The Male-Specific Region of the Human Y Chromosome Is a Mosaic of Discrete Sequence Classes." *Nature* 423: 825–837.

Snowden-Hill, D. P., E. S. Freeman, and P. B. Price. 1995. "Limits on Dark Matter Using Ancient Mica." *Physical Review Letters* 74:4133–36, and 76:332.

Sobel, Dava. 1995. *Longitude: The True Story of a Lone Genius Who Solved the Greatest Scientific Problem of His Time.* New York: Penguin.

Sorabji, Richard. 1987. *Philoponus and the Rejection of Aristotelian Science.* Ithaca, N.Y.: Cornell University Press.

Spanos, Nicholas P. 1996. *Multiple Identities and False Memories.* Washington, D.C.: American Psychological Association.

Spencer, Robert. 2002. *Islam Unveiled: Disturbing Questions About the World's Fastest-Growing Faith.* San Francisco: Encounter Books.

Stanton, Graham. 2002. *The Gospels and Jesus,* 2nd ed. Oxford: Oxford University Press.

Stark, Rodney. 1992. *The Churching of America, 1776–1990: Winners and Losers in Our Religious Economy.* New Brunswick, N.J.: Rutgers University Press.

Stark, Rodney. 1996. *Rise of Christianity: A Sociologist Reconsiders History.* Princeton: Princeton University Press.

Stark, Rodney. 1999. "Secularization, R.I.P." *Sociology of Religion* 60:249–73.

Stark, Rodney. 2001. *One True God: Historical Consequences of Monotheism.* Princeton: Princeton University Press.

Stark, Rodney. 2003. *For the Glory of God: How Monotheism Led to Reformations, Science, Witch-Hunts, and the End of Slavery.* Princeton: Princeton University Press.

Stephenson, F. R., and D. A. Green. 2002. *Historical Supernovae and Their Remnants.* Oxford: Clarendon Press.

Stonehouse, Julia. 1994. *Idols to Incubators: Reproduction Theory Through the Ages.* London: Scarlet Press.

Straton, John Roach. 1924. *The Famous New York Fundamentalist-Modernist Debates: The Orthodox Side.* New York: Doran.

Suzuki, Y., et al. 2000. "Localization of the Sex-Determining Region-Y Gene in XX Males." *Archives of Andrology* 44 (2): 133–36.

Swinburne, Richard. 1970. *The Concept of Miracle.* London: Macmillan.

Tavris, Carol. 2002. "The High Cost of Skepticism." *Skeptical Inquirer* 24 (4): 41–44.

Theissen, Gerd, and Annette Merz. 1989. *The Historical Jesus: A Comprehensive Guide.* London: SCM Press.

't Hooft, Gerard. 1976. "Symmetry Breaking Through Bell-Jackiw Anomalies." *Physical Review Letters* 37:8–11.

Tipler, Frank J. 1994a. *The Physics of Immortality.* New York: Doubleday.

Tipler, Frank J. 1994b. "God in the Equations." *Nature* 369:198.

Tipler, Frank J. 1996a. "Traveling to the Other Side of the Universe." *Journal of the British Interplanetary Society* 49:313–18.

Tipler, Frank J. 1996b. "Newtonian Cosmology Revisited." *Monthly Notices of the Royal Astronomical Society* 282:206–10.

Tipler, Frank J. 2003. "Intelligent Life in Cosmology." *International Journal of Astrobiology* 2:141–48.

Tipler, Frank J. 2005a. "Structure of the World from Pure Numbers." *Reports on Progress in Physics* 68:897–964.

Tipler, Frank J. 2005b. "The Star of Bethlehem: A Type Ia/Ic Supernova in the Andromeda Galaxy?" *Observatory* 125:168–73.

Tipler, Frank J. 2007. "The Value/Fact Distinction and Coase's Theorem." Preprint.

Tipler, Frank J., C. J. S. Clarke, and G. F. R. Ellis. 1980. "Singularities and Horizons: A Review Article." In *General Relativity and Gravitation: One Hundred Years After the Birth of Albert Einstein,* ed. A. Held, vol. 2, 97–206. New York: Plenum Press.

Tipler, Frank J., et al. 2007. "Closed Universes with Black Holes but No Event Horizons as a Solution to the Black Hole Information Problem." *Physical Review D.*

Trevor-Roper, H. R. 1967. *Religion, the Reformation and Social Change.* London: Macmillan.

Van den Bergh, Sidney. 2002. "The Light Curve of S Andromedae." *Astronomical Journal* 123:2045–46.

Vermes, Geza. 1973. *Jesus the Jew: A Historian's Reading of the Gospels.* Philadelphia: Fortress Press.

Vermes, Geza. 2000. *The Changing Faces of Jesus.* New York: Viking Compass.

Warner, Marina. 1976. *Alone of All Her Sex: The Myth and Cult of the Virgin Mary.* London: Picador.

Weigel, George. 2002. *The Courage to Be Catholic.* New York: Basic Books.

Weinberg, Steven. 1996. *The Quantum Theory of Fields,* vol. 2. Cambridge: Cambridge University Press.

Weinberg, Steven. 2001. *Facing Up: Science and Its Cultural Adversaries.* Cambridge, Mass.: Harvard University Press.

Weston, Jessie L., trans. 1904. *Sir Gawain at the Grail Castle.* Repr., Dyfed, Wales: Llanerch Publishers, 1995.

Weston, Jessie L. 1913. *The Quest of the Holy Grail.* London: G. Bell and Sons. Repr., New York: Dover Publications, 2001.

Wilczek, Frank. 2002. "Scaling Mount Planck III: Is That All There Is?" *Physics Today,* August, 10–11.

Wilson, Amy 2002. "War and Remembrance." *Orange County Register,* November 3.

Wilson, Ian. 1998. *The Blood and the Shroud: New Evidence That the World's Most Sacred Relic Is Real.* New York: Simon & Schuster.

Wilson, Ian, and Barrie Schwortz. 2000. *The Turin Shroud: The Illustrated Evidence.* New York: Barnes & Noble Books.

Winston, N., et al. 1991. "Parthenogenetic Activation and Development of Fresh and Aged Human Oocytes." *Fertility and Sterility* 53:266–79.

Wright, N. Thomas. 2003. *The Resurrection of the Son of God.* Minneapolis: Fortress Press.

Zel'dovich, Yacob B. 1972. "A Hypothesis Unifying the Structure and the Entropy of the Universe." *Monthly Notices of the Royal Astronomy Society* 160:1P–3P.

Zenteno, J. C., et al. 1997. "Two SRY-Negative XX Brothers Without Genetal Ambiguity." *Human Genetics* 100:606–10.

Zhang, Y. W. 2000. "Identification of Monozygotic Twin Chimpanzees by Microsatellite Analysis." *American Journal of Primatology* 52:101–106.

Ziegler, J. F., J. P. Biersack, and U. Littmark. 1985. *The Stopping and Range of Ions in Solids.* New York: Pergamon.

Zirkle, Conway. 1936. "Animals Impregnated by the Wind." *Isis* 25 (1):95–130.

Index

Page numbers in *italics* refer to figures, illustrations, and tables.

INDEX

Art Credits

Figure 5.1: Painting © Philadelphia Museum of Art; purchased with the W.P. Wilstach Fund, 1899. Used with permission.

Figure 6.1: Photograph © Ill Schoening and Venessa Harvey, REU Program/National Optical Astronomy Observatory/Association of Universities for Research in Astronomy/National Science Foundation.

Figure 6.2: Eclipse map courtesy of Fred Espenak, NASA/Goddard Space Flight Center.

Figure 7.1: Photograph © Barrie Schwortz, 1978. Used by permission.

Figure 7.2: Photograph © Mark Evans, 1978. Used by permission.

Figure 7.3: Photograph © Centro Español de Sindologia. Reproduced with kind permission.

Figure 8.1: Photograph © Mark Evans, 1978. Used by permission.

Figure 8.2: Photograph © by Dr. Leen Ritmeyer. Used by permission.

Figure 8.3: Photograph © by Dr. Leen Ritmeyer. Used by permission.

Figure 8.4: Photograph © Robert L. Fleischer. Used by permission.

Figure 8.5: Photograph © Robert L. Fleischer. Used by permission.

Figure 8.6: Drawing of Jerusalem © Dr. Leen Ritmeyer. Used by permission.

Courtesy of the author

Frank J. Tipler is a professor of mathematical physics at Tulane University and the author of *The Physics of Immortality*. He lives in New Orleans, Louisiana.